FIRST EDITION

MICROPROCESSOR INTERFACE

BY **CHUNG S. LEUNG**
TEXAS A&M UNIVERSITY-KINGSVILLE

Bassim Hamadeh, CEO and Publisher
Kassie Graves, Director of Acquisitions
Jamie Giganti, Senior Managing Editor
Miguel Macias, Senior Graphic Designer
Marissa Applegate, Senior Field Acquisitions Editor
Gem Rabanera, Project Editor
Elizabeth Rowe, Licensing Coordinator
Allie Kiekhofer, Associate Editor
Kat Ragudos, Interior Designer

Copyright © 2017 by Cognella, Inc. All rights reserved. No part of this publication may be reprinted, reproduced, transmitted, or utilized in any form or by any electronic, mechanical, or other means, now known or hereafter invented, including photocopying, microfilming, and recording, or in any information retrieval system without the written permission of Cognella, Inc.

Trademark Notice: Product or corporate names may be trademarks or registered trademarks, and are used only for identification and explanation without intent to infringe.

Cover image copyright © Depositphotos/kilukilu.

Printed in the United States of America

ISBN: 978-1-5165-0555-5 (pbk) / 978-1-5165-0556-2 (br)

CONTENTS

	PREFACE	1
	ACKNOWLEDGMENTS	3
PART I	**Microprocessor Interface Projects**	5
	HOW TO WORK WITH CHARACTER-BASED LCDS	6
	LCD ADDRESSING FOR THE MOST COMMON LCD	8
	HOW TO MAKE A CHARACTER-BASED LCD TO DISPLAY THE DATA	13
	PROJECT #1: DIGITAL THERMOMETER	15
	PROJECT #2: ELECTRONIC TAPE MEASURE	17
	The accuracy evaluation of your electronic tape measure	17
	Linearize a GP2Y0A02 IR sensor with an 8-bit ATD conversion	18
	How to modify the following MATLAB file with your calibrated data	21
	PROJECT #3: MEASURE THE ROTATIONS PER MINUTE OF A BRUSHLESS DC MOTOR	33
	PROJECT #4: LCD TIMER/STOPWATCH	35
	Pros and cons of ASCII code and the decimal increment display	38
	How to convert this timer to a stopwatch	39

PROJECT #5: IR TRIGGERING CONTROL FOR TIME ELAPSE — 41

Another triggering control method with FSR — 42

PROJECT #6: LCD DIGITAL CLOCK (MILITARY TIME) WITH ALARM — 43

PROJECT #7: LCD DIGITAL CLOCK WITH ALARM (AM/PM) — 47

PROJECT #8: HOME SECURITY SYSTEM — 49

Using the {} key* — 51

Arming commands with {2} or {3} keys — 51

Quick arming — 52

Green/red LED — 52

Chime mode, {9} key — 52

Using the {OFF} or {1} key — 53

Using the {BYPASS} or {6} key — 53

PROJECT #9: SECONDARY PIN FOR THE HOME SECURITY SYSTEM — 55

PART II Autonomous Robots — 57

INTRODUCTION TO SERVO MOTORS — 58

HOW TO CONTROL A SERVO MOTOR — 59

HOW TO CHOOSE A CHASSIS — 61

PROJECT #10: CIRCULAR MOTION — 63

PROJECT #11: FIGURE EIGHT CONFIGURATION — 65

PROJECT #12: OBSTACLE AVOIDANCE — 67

PROJECT #13: NAVIGATE OUT OF A DEAD END — 69

PROJECT #14: RUNNING THE MAZE — 71

PART III Microprocessor Labs — 73

- **LAB #1:** ADDRESSING MODE — 75
- **LAB #2:** COMPUTE THE ARRAY SUM — 79
- **LAB #3:** CONDITION FLAGS AND ROTATE INSTRUCTION — 83
- **LAB #4:** BIT TESTING — 89
- **LAB #5:** BIT MANIPULATION — 91
- **LAB #6:** LED TRAFFIC LIGHT — 93
- **LAB #7:** PIN VERIFICATION WITH 4X4 MATRIX KEYPAD — 95
- **LAB #8:** PIN VERIFICATION WITH SINGLE POLE/COMMON BUS KEYPAD — 97
- **LAB #9:** SIREN GENERATION — 103
- **LAB #10:** DIM AN LED — 105
- **LAB #11:** HOME SECURITY SYSTEM — 107

PREFACE

There are many texts on microprocessors, but trying to locate an idea for a microprocessor interface project would be quite a challenge. Most of the ideas are scattered here and there all over the Internet. Trying to implement some of those ideas will take creativity, because most of the modules are not available. That is the reason why I have had to pull out my tools from my garage and built the modules myself for the course.

The purpose in writing this text is to share what I have been asking my students to do—design and gain hands-on experience with microprocessors. This text is divided into three parts: Part I covers ideas in microprocessor interface; Part II covers essential designed components in an autonomous robot; and Part III covers the essential ideas for the fundamental components of microprocessor systems. Most of the time students are learning the programming from examples. As most of the projects will take at least two to three weeks to complete, it is almost impossible to finish more than five projects in a regular fifteen-week semester. It will be beneficial to cover some of the essential components, like creating a two-tone siren in a microprocessor lab; then, when students are in the microprocessor interface, they will be equipped with the knowledge to activate the buzzer when the temperature sensor reaches steady body temperature. Of course, building an interface project has to be an individual endeavor. If they are working in a team of three, then most likely one of them will be a benchwarmer or cheerleader. My philosophy is to advise students to keep talking to the development board until it talks back to them. The solution manual is embedded inside the microprocessor development board.

I hear and I forget.   I see and I remember.   I do and I understand.

— Confucius

Some of the project ideas are related to each other. For example, when a student can come up his or her version of a digital stopwatch, then they can utilize the stopwatch to determine the time lapse for an object to roll down the slider. Of course, someone has to take the initiative to build the slider—not just one but about ten sliders for a class of twenty students. All the modules that I have built could probably fit in a backyard shed.

Part II is dedicated to autonomous robots. There are various autonomous robotic competitions—even high school students are using Lego kits to get into robotics. As for university students, most of them are using Arduino development kits to implement their robots.

2 MICROPROCESSOR INTERFACE

The most important step for a thousand-mile journey is the first step! Most of the time students overlook this first step.

What kind of robot do you want to build?
What features and how much memory does your robot need?
How many input and output pins you will need in your development board?

If you need to have a camera on board, then you will need to look into the development board that can interface with a camera. The lowest resolution for any camera is 400 by 600 pixels, 400×600 = 240,000 pixels per picture. Assuming you are taking black and white pictures only, each pixel will be represented by at least 1 byte. With a simple calculation, you can figure out that it will take 240 KB to store a 400×600 picture. The ultimate question will be, how much memory will you need on your development board? 512 KB or 1 MB? Remember your program takes up some memory space, too!

In general, it is much easier to have a sensor to identify the color of an object than the shape of the object. Using MATLAB, you can do the pattern recognition with your eyes, but trying to program a microprocessor to do the pattern recognition can be quite a task!

The main purpose in appending the robotics in this text is to provide certain insights into some of the basic features of an autonomous robot. Every year, I have many students come to my office and ask me how to drive a servo, so the servo is introduced in the beginning of Part II in this text. I have seen student teams spend several months in designing their autonomous robot for competition, and then realize that they need to have camera interface with the development board. Starting the process over in the middle of the semester is almost like changing horses in the middle of a stream. This type of mistake can be quite costly! On the other hand, I have also seen a student team garnish their robot with eight infrared or IR sensors, and actually none of them were working correctly. Last but not least, one student team designed an autonomous robot with a fifth-degree of freedom robotic arm. The robotic arm was designed with a 3-D printer, but the material they used was too heavy for the servos. They had to synchronize two servo motors (Hercules) to work as one, and eventually the weight of the robot arm slowed everything down to a crawl. They could have replaced the robotic arm with a gripper, lifted the object off the ground by five degrees and moved the object to the designated location. There are many ways to accomplish a goal, so as long as the student team can figure out the pros and cons for each approach, they can reach the best approach to achieve their goal.

Part III illustrates some of the lab components for my Microprocessor Systems class. We are using Huang's HCS12/9S12 text: *An Introduction to Software and Hardware Interfacing*. Some of the programs for the lab can be found in Huang's text, and we are using Dragon12-JR from EVBplus (Wytec Company). Although EVBplus had told me that only a few universities are using Dragon12-JR for their microprocessor class, I still believe this development board is perfect to mount on the chassis kits from budgetrobotics.com. If there is interest in pursuing any autonomous robot, then the instructor can change the Lab #10 in Part III (Dim an LED) to run the servo motor with pulse width modulation. If the instructor asks students to design a home security system, then the Security System Lab is a must for the microprocessor lab; many students have no idea how to monitor two doors in one zone. Lab #9 in Part III: Siren Generation will help students to develop a basic idea of how to sound a buzzer when digital thermometer reaches body temperature. Since the title of the text is Microprocessor Interface, it is more appropriated to arrange the microprocessor interface projects first, the autonomous robot second, and the laboratory components for the microprocessor systems last.

ACKNOWLEDGMENTS

I want to thank all of the electrical engineering students who took the Microprocessor Systems, Advanced Lab, and Microprocessor Interface classes over the past ten years at Texas A&M University–Kingsville, as they contributed many good ideas for this text. Particularly, I am grateful for four of my students: Jose Lerma, Christopher Pefanis, Eric Wineman, and Wolfgang Schwertner. They have contributed many hours to implementing the microprocessor interface projects, autonomous robots, and microprocessor labs. Special thanks to one mechanical engineering student, Agnes Su, who drafted out all the isometric views of the setup for the projects.

I am also grateful for Hong Kong Philips Electronics Company, in collaboration with the Hong Kong Labor Department, for providing me with five-year apprenticeship training right after I finished high school. This hands-on training molded me into a skillful, hands-on engineer. My teaching Microprocessor Systems, Microprocessor Interface, and Senior Project was recommended by one of my students to the Chair more than twelve years ago. I have been teaching these courses since then, and enjoy every minute of it!

Last but not least, I have to thank my parents, for their love and trust in me. Without their sacrifice and faith, I could not have achieved what I have now.

PART I

MICROPROCESSOR INTERFACE PROJECTS

HOW TO WORK WITH CHARACTER-BASED LCDS

LCD ADDRESSING FOR THE MOST COMMON LCD

HOW TO MAKE A CHARACTER-BASED LCD TO DISPLAY THE DATA

PROJECT #1 *Digital Thermometer*

PROJECT #2 *Electronic Tape Measure*

PROJECT #3 *Measure the Rotations per Minute of a Direct Current Motor*

PROJECT #4 *LCD Timer/Stopwatch*

PROJECT #5 *IR Triggering Control for Time Elapse*

PROJECT #6 *LCD Digital Clock (Military Time) with Alarm*

PROJECT #7 *LCD Digital Clock with Alarm (AM/PM)*

PROJECT #8 *Home Security System*

HOW TO WORK WITH CHARACTER-BASED LCDS

Most character-based Liquid Crystal Display (LCD) modules use a Hitachi HD44780 (or compatible) controller chip. These modules are not as sophisticated as the latest full-size, full-color, back-lit types used in laptop computers. The character-based LCDs find their way in simple display requirements. For educational purposes, we will utilize the LCD extensively for displaying data values.

Nowadays we mostly use microprocessor development boards for teaching microprocessor classes. Most development boards will have a built-in LCD port. Students will have to get a ribbon cable and connect the LCD to the LCD port. In general, the LCD pin has two formats: a single row with fourteen pins or two rows with seven pins in each row. The two layouts are as shown in Figure I.1.

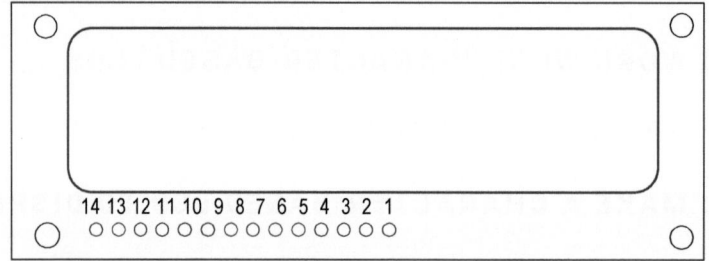

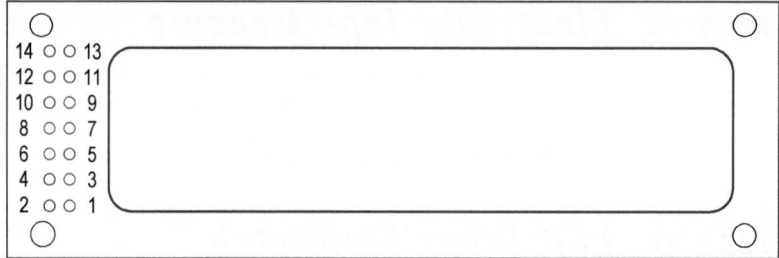

FIGURE I.1: Common pin configuration for the LCD

TABLE I.1: The command control for a character-based LCD

Command	Binary								HEX
	D7	D6	D5	D4	D3	D2	D1	D0	
Clear display	0	0	0	0	0	0	0	1	$01
Display and cursor home	0	0	0	0	0	0	1	x	$02/03
Character entry mode	0	0	0	0	0	1	I/D	S	$04~07
Display on/off and cursor	0	0	0	0	1	D	C	B	$08~0F
Display/cursor shift	0	0	0	1	S/C	R/L	x	x	$10~1F
Function set	0	0	1	DL	N	F	x	x	$20~3F

Set CGRAM address		0	1	CGRAM address		$40~7F
Set display address			1	DDRAM address		$80~B7
	Bit value = 1			**Bit value = 0**		
I/D	Increment *			Decrement		
S	Display shift ON			Display shift OFF *		
D	Display ON			Display OFF *		
C	Cursor ON			Cursor OFF *		
B	Cursor blink ON			Cursor blink OFF *		
S/C	Shift display			Cursor move		
R/L	Right shift			Left shift		
DL	8-bit interface *			4-bit interface		
N	2-line mode			1-line mode *		
F	5 × 10 dot			5 × 7 dot *		

Note: x = don't care; * = initialization settings.

Set 4-bit data format, 2-line display, and 5 × 7 font:

Command	D7	D6	D5	D4	D3	D2	D1	D0	HEX
Function set	0	0	1	0	1	0	0	0	$28

DL (D4):   0 = 4-bit interface *   N (D3):   1 = 2-line display   F (D2): 0 = 5 × 7 dot *

Turn on display, with the cursor underline and blinking:

Command	D7	D6	D5	D4	D3	D2	D1	D0	HEX
Display on/off and cursor	0	0	0	0	1	1	1	1	$0F

D (D2):   1 = Display ON   C (D1):   1 = Cursor ON   B (D0): 1 = Cursor blink ON

Move cursor to the right for entry mode:

Command	D7	D6	D5	D4	D3	D2	D1	D0	HEX
Character entry mode	0	0	0	0	0	1	1	0	$06

I/D (D1):   1 = Increment *   S (D0):   0 = Display shift OFF

LCD ADDRESSING FOR THE MOST COMMON LCD

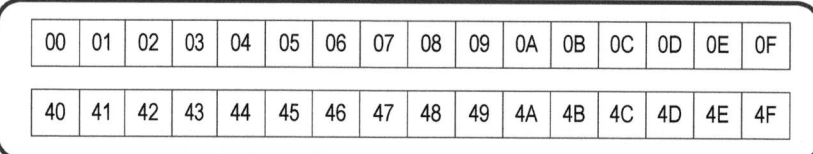

FIGURE I.2: Addresses in hex and display locations for 2×16 LCD

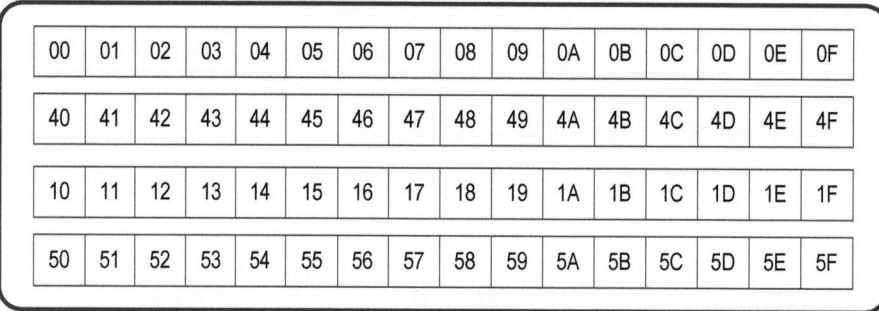

FIGURE I.3: Addresses in hex and display locations for 4×16 LCD

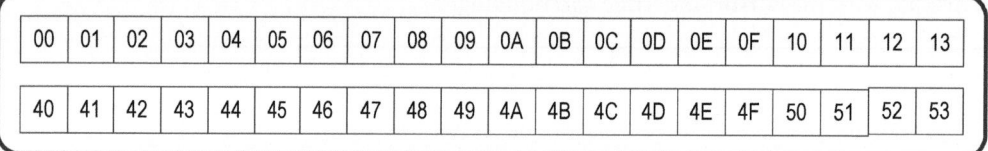

FIGURE I.4: Addresses in hex and display locations for 2×20 LCD

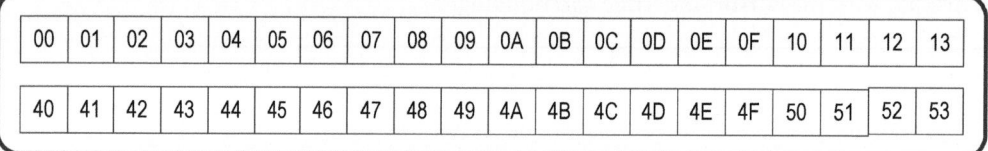

FIGURE I.5: Addresses in hex and display locations for 4×20 LCD

More illustrations can be found at http://web.alfredstate.edu/weimandn/lcd/lcd_addressing/lcd_addressing_index.html.

```
#include "Reg9s12.h"              ; define the path for your header file
LCD_dat     equ     PortK         ; LCD data pins (PK5-PK2)
LCD_dir     equ     DDRK          ; LCD data direction port
LCD_E       equ     $02           ; E signal pin
LCD_RS      equ     $01           ; RS signal pin

            org     $2000
            lds     #$2000        ; set up stack pointer
            jsr     openLCD       ; initialize LCD
            ldaa    #$82
            jsr     cmd2LCD
            ldx     #msg1         ; msg1 = "HELLO WORLD"
            jsr     puts2LCD
            ldaa    #$C0          ; move to the second row
            jsr     cmd2LCD

            ldx     #msg2         ; msg2 = "LCD IS WORKING!"
            jsr     puts2LCD
            swi

msg1        fcc     "HELLO WORLD"
            dc.b    0
msg2        fcc     "LCD IS WORKING!"
            dc.b    0

openLCD     movb    #$FF,LCD_dir  ; configure Port K for output
            ldy     #10           ; wait for LCD to be ready
            jsr     delay10ms     ;          "
            ldaa    #$28          ; set 4-bit data, 2-line display, 5x8 font
            jsr     cmd2LCD       ;          "
            ldaa    #$0F          ; turn on display, cursor, and blinking
            jsr     cmd2LCD       ;          "
            ldaa    #$06          ; move cursor right
            jsr     cmd2LCD       ;          "
            ldaa    #$01          ; clear screen and return to home position
            jsr     cmd2LCD       ;          "
            ldy     #2            ; wait until clear display command is complete
            jsr     delayby1ms    ;          "
            rts

cmd2LCD     psha                  ; save the command in stack
            bclr    LCD_dat,LCD_RS ; select the instruction register
            bset    LCD_dat,LCD_E ; pull the E signal high
            anda    #$F0          ; clear the lower 4 bits
            lsra                  ; match the upper 4 bits with the LCD
```

```
          lsra                         ; data pins
          oraa    #LCD_E               ; maintain the E signal value
          staa    LCD_dat              ; send the command, RS, and E signals
          nop                          ; extend the duration of the E pulse
          nop                          ;         "
          nop                          ;         "
          bclr    LCD_dat,LCD_E        ; pull the E signal low
          pula                         ; retrieve the LCD command
          anda    #$0F                 ; clear the upper 4 bits
          lsla                         ; match the lower 4 bits with the LCD
          lsla                         ; data pins
          bset    LCD_dat,LCD_E        ; pull the E signal high
          oraa    #LCD_E               ; maintain the E signal value
          staa    LCD_dat              ; send the lower 4 bits with E and RS
          nop                          ; extend the duration of the E pulse
          nop                          ;         "
          nop                          ;         "
          bclr    LCD_dat,LCD_E        ; clear the E signal
          ldy     #1                   ; adding this delay will complete the
          jsr     delayby50us          ; internal operation for most instructions
          rts
putc2LCD  psha                         ; save a copy of the data
          bset    LCD_dat,LCD_RS       ; select LCD data register
          bset    LCD_dat,LCD_E        ; pull E to high
          anda    #$F0                 ; mask out the lower 4 bits
          lsra                         ; match the upper 4 bits with the LCD
          lsra                         ; data pins
          oraa    #$03                 ; keep signal E and RS unchanged
          staa    LCD_dat              ; send the upper 4 bits and E, RS signals
          nop                          ; provide enough duration to the E signal
          nop                          ;         "
          nop                          ;         "
          bclr    LCD_dat,LCD_E        ; pull the E signal low
          pula                         ; retrieve the character from the stack
          anda    #$0F                 ; clear the upper 4 bits
          lsla                         ; match the lower 4 bits with the LCD
          lsla                         ; data pins
          bset    LCD_dat,LCD_E        ; pull the E signal high
          oraa    #$03                 ; keep E and RS unchanged
          staa    LCD_dat
          nop
          nop
          nop
          bclr    LCD_dat,LCD_E        ; pull E low to complete the write cycle
          ldy     #1                   ; wait until the write operation is
          jsr     delayby50us          ; complete
```

```
                rts

puts2LCD        ldaa      1,x+              ; get one character from the string
                beq       done_puts         ; reach NULL character?
                Jsr       putc2LCD
                Bra       puts2LCD
done_puts       rts
;;;;;;;;;;;;;;;;;;;;;;;;;Time Delays;;;;;;;;;;;;;;;;;;;;;;;;;;;;
; the following subroutine creates a 50-us delay
delay50us       movb      #$90,TSCR1        ; enable TCNT & fast flag clear
                movb      #$01,TSCR2        ; configure prescale factor to 2
                bset      TIOS,$01
                ldd       TCNT              ; interference with LCD screen update
again0          addd      #600              ; start an output compare operation
                std       TC0               ; with 10-ms time delay
                brclr     TFLG1,$01,*
                ldd       TC0
                dbne      y,again0
                rts

; the following subroutine creates a 1-ms delay
delay1ms        movb      #$90,TSCR1        ; enable TCNT & fast flag clear
                movb      #$06,TSCR2        ; configure prescale factor to 64
                bset      TIOS,$01
                ldd       TCNT
again1          addd      #375              ; start an output compare operation
                std       TC0               ; with 1-ms time delay
                brclr     TFLG1,$01,*
                ldd       TC0
                dbne      y,again1
                rts

; the following subroutine creates a 10-ms delay
delay10ms       movb      #$90,TSCR1        ; enable TCNT & fast flag clear
                movb      #$06,TSCR2        ; configure prescale factor to 64
                bset      TIOS,$01
                ldd       TCNT
again2          addd      #3750             ; start an output compare operation
                std       TC0               ; with 10-ms time delay
                brclr     TFLG1,$01, *
                ldd       TC0
                dbne      y,again2
                rts
```

Note: With respect to the software AsmIDE, *Reg9s12.h* is the header file to map all the PINs or Ports to a certain address. If *Reg9s12.h* is not included in your program, the addresses for all those pins or ports have to be assigned individually in the beginning of the assembly program, which can take quite a bit of time. The original version of AsmIDE340 had PTA stands for PortA and PTB stands for PortB. PTA and PTB have been replaced with PortA and PortB for clarification purposes in my version of *Reg9s12.h*. If there are errors running the above program related to PortA or PortB in your version of AsmIDE that is the reason why.

The LCD output format for the above assembly program is as shown below:

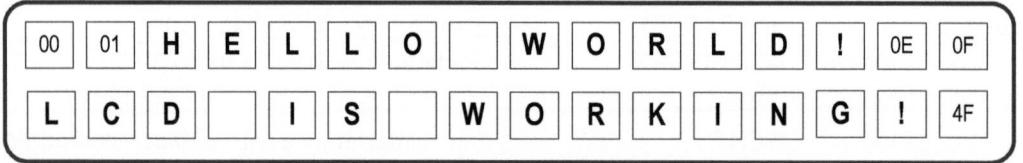

FIGURE I.6: LCD output format

With reference to the instructions in the beginning of the assembly program:

```
         jsr      openLCD           ; initialize LCD
         ldaa     #$82              ; DDRAM address is @ $02
         jsr      cmd2LCD
         ldx      #msg1             ; msg1 = "HELLO WORLD!"
```

According to the set display address in Table I.2a:

TABLE I.2A

Command	Binary								HEX
	D7	D6	D5	D4	D3	D2	D1	D0	
Set display address	1	0	0	0	0	0	1	0	$82
	1	DDRAM address = $02							$82

The address $02 (where H is located) is the starting address for the first character on the first row of the LCD. If you remove the instruction ldaa #$82 after jsr openLCD, then DDRAM address is initialized as $00.

With reference to the instructions for msg2 in the beginning of the assembly program:

```
         ldaa     #$C0              ; move to the second row
         jsr      cmd2LCD
         ldx      #msg2             ; msg2 = "LCD IS WORKING!"
```

According to the set display address in Table I.2b:

TABLE I.2B

Command	Binary								HEX
	D7	D6	D5	D4	D3	D2	D1	D0	
Set display address	1	1	0	0	0	0	0	0	$C0
	1	DDRAM address = $40							$C0

The address $40 (where L is located) is the starting address for the first character on the second row of the LCD.

The above assembly program is using 4-bit data interfacing with the LCD, since the majority of microprocessor systems classes are using development boards to learn how to program the microprocessor. On each of the development boards, there should be an LCD socket for connecting your LCD with a ribbon cable. Before you attempt to use 8-bit data interfacing with the LCD, please review the configuration of your development board. Some of the development boards might not have the connections built in for 8-bit data interfacing with the LCD; in that case you have to make your own connections with the jumper wires.

HOW TO MAKE A CHARACTER-BASED LCD TO DISPLAY THE DATA

Throughout several projects in this text, you are required to display the data value on the LCD; for example, in project #1 you are asked to display the temperature on the LCD. According to the data sheet of LM34, when the output of the circuit is at 750 mV, it is equivalent to 75°F. Of course, this output voltage has to be converted by means of the analog-to-digital (ATD) converter in the development board, and the digitized value will be compared with the digitized value for 700 mV and 800 mV. Divide this value by 100, and if the answer is equal to 1 then the most significant digit should be 1. If the answer is 0, then the most significant digit can be skipped. The remainder will be divided by 10, and if the answer is 0, the second digit should be 0; if the answer is 7, then the second digit should be 7. The three digits before the decimal and the digit after the decimal can all be determined in a similar fashion. In order to display the digit 7 on the LCD, you have to send the ASCII code for digit 7 ($37) to the LCD data path. Once you accomplish this comparison routine, you can use it for the electronic tape measure project to display the distance—except a linear equation has to be determined.

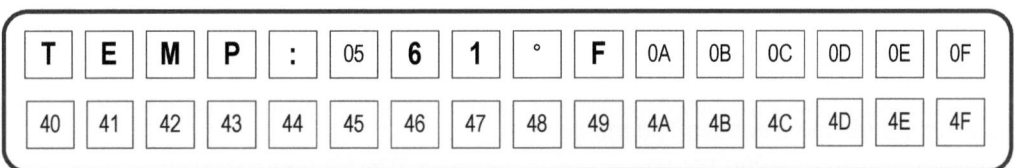

FIGURE I.7: LCD temperature output format

The output LCD format for the following assembly program, including those four LCD functions and the delay functions, is as shown above. The digit 1 in DDRAM address $07 will refresh with digit 2 in two seconds, then digit 3, digit 4, and digit 5. In your project, the ASCII code of your data value from the ATD converter will be fed to the data path.

```
#include     "Reg9s12.h"        ; define the path for your header file
LCD_dat      equ     PortK      ; LCD data pins (PK5-PK2)
LCD_dir      equ     DDRK       ; LCD data direction port
LCD_E        equ     $02        ; E signal pin
```

```
LCD_RS       equ      $01                  ; RS signal pin

             org      $2000
             lds      #$2000               ; set up stack pointer
             jsr      openLCD              ; initial LCD
             ldx      #msg                 ; msg = "Temp:"
             jsr      puts2LCD
             ldx      #5                   ; set loop counter to 5
             ldab     #$30                 ; set initial value to 0 (ASCII = $30)
             pshb                          ; save it onto stack
             ldaa     #$86                 ; DDRAM address = $06
             jsr      cmd2LCD
             ldaa     #$36                 ; most significant character is 6 (ASCII = $36)
             jsr      putc2LCD
             ldaa     #$89                 ; DDRAM address = $09
             jsr      cmd2LCD
             ldaa     #$46                 ; ASCII code for upper case F
             jsr      putc2LCD
loop         ldaa     #$87                 ; DDRAM address = $07
             jsr      cmd2LCD
             pulb                          ; pull the ASCII code from the stack to B
             incb                          ; increment the ASCII code by 1
             tba                           ; transfer from B to A
             pshb                          ; save it onto the stack
             jsr      putc2LCD
             ldy      #20                  ; looping the delay100ms for 20 times
             jsr      delay100ms           ; to implement the 2 second delay
             dbne     x,loop
             swi
msg          fcc      "Temp:"
             dc.b

; the following subroutine creates a 100 ms delay
delay100ms   movb     #$90,TSCR1           ; enable TCNT & fast flag clear
             movb     #$06,TSCR2           ; configure prescale factor to 64
             bset     TIOS,$01
             ldd      TCNT
again3       addd     #37500               ; start an output compare operation
             std      TC0                  ; with 10 ms time delay
             brclr    TFLG1,$01,*
             ldd      TC0
             dbne     y,again3
             rts
```

PROJECT #1:
Digital Thermometer

 For Fahrenheit, you can use LM 34

http://www.ti.com/lit/ds/symlink/lm34.pdf

 For Celsius, you can use LM 35

http://www.ti.com/lit/ds/symlink/lm35.pdf

Write an assembly program to interface the temperature sensor LM34/LM35 with the microcontroller. This will be a single channel scanning. Try to slow down the ATD (analog-to-digital) clock as much as possible. (The range of the ATD clock can be set between 500 KHz ~ 2 MHz.) The ASCII code for the degree symbol is ° ⌧ $B0.

Each conversion should be saved in the result registers. Allow the temperature sensor to rise to your body temperature by holding the temperature sensor with your fingers. This might not be sufficient for reaching your body temperature (98.6°F or 37°C). If the grip on the temperature sensor is firm enough, it will eventually reach body temperature. Remember to set the V_{RH} and V_{RL} for the ATD. Increase the time span to a longer period if necessary. When the data in the data register stays constant for eight or ten samples/conversions, sound the buzzer twice (same as the digital thermometer in the market). Of course, the temperature will be displayed on the LCD as the temperature sensor reaches a certain temperature periodically until it reaches a steady state.

The LM34/35 temperature sensor maps the temperature linearly with the output voltage. For example, when the temperature is 76°F, the voltage output for the LM34 should be around 760 mV. If the V_{RH} and V_{RL} for the ATD are not set correctly, your temperature display on the LCD will be off from 76°F. The temperature display on the LCD should be able to show three digits ahead of the decimal and one digit after the decimal in Fahrenheit. If the temperature display on the LCD is in centigrade, there should be at least two digits ahead of the decimal and one digit after the decimal. The program should convert the binary output of the ATD back into millivolts. The following equation can be used for the 10-bit conversion:

$$\text{Decimal value to LCD in } mV = \text{10-bit binary converted to decimal} \times \frac{5000}{2^{10}-1}$$

Note: V_{RH} is equal to 5 V.

Since the LCD is to display a character, a routine has to convert the output voltage into a character one at a time. The ATD converts each sample of the body temperature into binary data and stores it in the result data register. First, read the binary data from the result data register, and convert it to a decimal (e.g., 765 mV is equivalent to 76.5°F, and 890 mV is equivalent to 89°F, etc.). Divide this value by 100, and if the answer is equal to 1 then the most significant digit should be 1. If the answer is 0, then the most significant digit can be skipped. The remainder will be divided by 10, and if the answer is 0, the second digit should be 0; if the answer is 7, then the second digit should be 7. The three digits before the decimal and the digit after the decimal can all be determined in a similar fashion.

Note: It is more reliable to use the power adaptor for the Dragon12-JR. If your development board draws the power from your laptop, it may not be fully charged even if you plug in the power adaptor for your laptop. The ATD in your development board definitely requires a full 5 V to provide an accurate digital output.

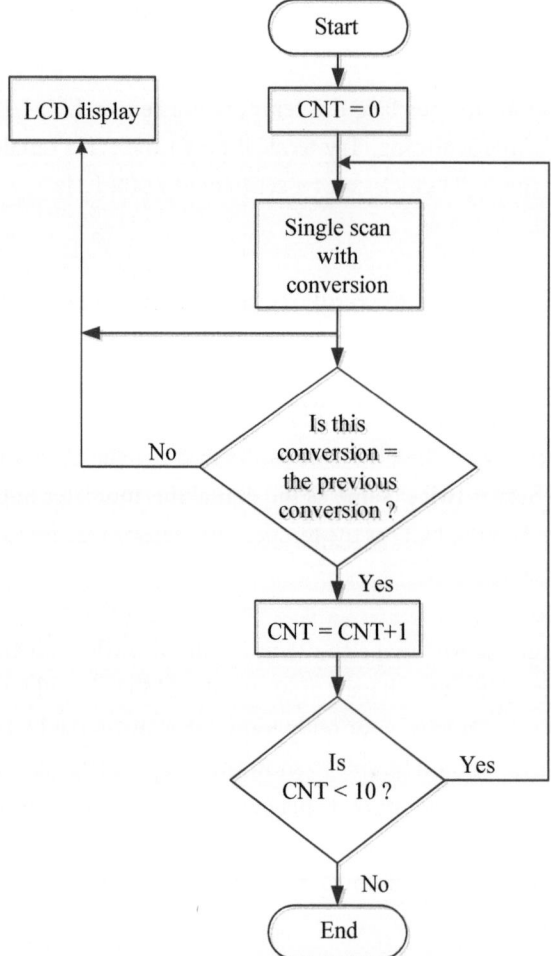

FIGURE 1.1: Flowchart for digital thermometer

PROJECT #2:
Electronic Tape Measure

Interface the 9S12 with the IR sensor (Sharp GP2Y0A02) and display the output on LCD. This Sharp IR sensor can measure any range between 20~150 centimeters or 8~60 inches. The final product should show the detected distance on the LCD in centimeters and in inches, with accuracy up to a quarter of an inch. Be sure to have the IR sensor working all the time as soon as the unit is ON.

Each IR sensor's characteristic curve is slightly different. You have to derive the linear function for your IR sensor, and use the same IR sensor to measure the distance. The linearization of an IR sensor corresponds to an 8-bit ATD conversion for Sharp GP2D12 IR sensor.[1] Follow a similar procedure to derive a linear function for GP2Y0A02.

Tape a tape measure on the bench, and align the zero against the wall or the obstacle panel. Move all objects and equipment away from the wall or obstacle panel. Then move the IR sensor away from the wall or panel and measure the received voltage corresponding to the distance away from the wall. The distance from the wall to the bracket with the IR sensor, minus a quarter inch, is the actual distance between the wall and the IR sensor. It is preferable to collect all the data without any natural light shining on the IR sensor. Type in all the data in a spreadsheet, and optimize the linear function. A common mistake is not being able to show one digit after the decimal. The remedy is to fine-tune the optimized function. Since the characteristic between the distance and the received voltage is non-linear, trial and error for a different constant will not work for this scenario.

THE ACCURACY EVALUATION OF YOUR ELECTRONIC TAPE MEASURE

Repeat all the data points for each distance measurement for your spreadsheet program, and record the distance displayed on your LCD. Use the spreadsheet to compute the square of each offset, and evaluate the mean squared error for all the data points.

---

[1] *http://www.acroname.com/robotics/info/articles/irlinear/irlinear.html#e4*

TABLE 2.1: Cross reference of most common Sharp IR sensor

Part number	Object detection range (cm)	Object detection range (in)
GP2D12	10 ~ 80	4 ~ 30
GP2D15	Preset range = 24	Preset range = 9.5
GP2D120	4 ~ 30	1.6 ~ 12
GP2Y0A02YK	20 ~ 150	8 ~ 60
GP2Y0A21YK	10 ~ 80	4 ~ 30
GP2Y0A710YK	100 ~ 550	40 ~ 216
GP2Y0D02YK	Preset range = 80	Preset range = 30
IS471F	5.1 ~ 7.6	2 ~ 3

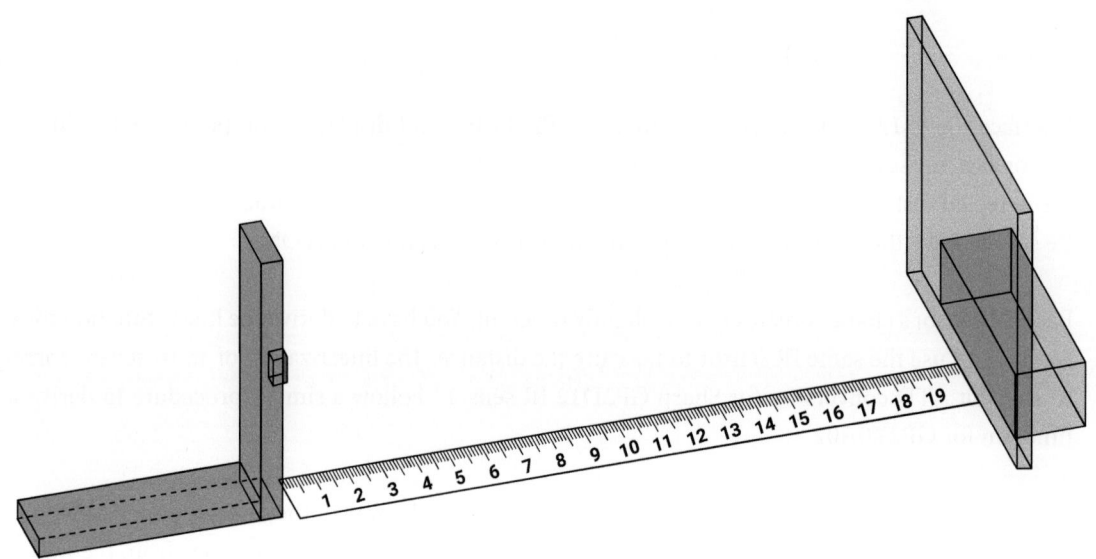

FIGURE 2.1: Electronic tape measure setup

LINEARIZE A GP2Y0A02 IR SENSOR WITH AN 8-BIT ATD CONVERSION

- With the illustration as shown in Figure 2.1, use a multi-meter to measure the rebounded signal voltage from the wall/panel. The signal voltage may not be in integers; make sure you record the digits after the decimal. At the same time, record the content of the ATD data register. *Note:* The data register only holds integer value.

- An ATD converter needs a low reference voltage (V_{RL}) and a high reference voltage (V_{RH}) to perform the conversion. The V_{RL} is often tied to ground, and the V_{RH} is often set to V_{DD} (supply voltage for ATD). Table 2.2 is the illustration for an 8-bit number in an ATD data register corresponding to an analog voltage. The ATD conversion result k corresponds to an analog voltage V_k, which is given by the following equation:

- $$V_K = V_{RL} + \frac{(V_{RH} - V_{RL}) \times k}{2^N - 1} = \frac{5}{2^8 - 1} \times k = \frac{5 \times k}{255} = \frac{k}{51}$$

where

$$V_{RL} = 0\ V,\ V_{RH} = 5\ V, \quad \text{and } N = 8 \text{ (for 8-bit ATD)}.$$

- The 8-bit representation in the ATD data register is an approximation of the rebounded signal voltage. For example, with 2.8 V, corresponding to Table 2.2, the ATD conversion value should be between 142 and 143, but the format of the 8-bit ATD gives a value of 143 instead of 142.801. Certain quantization error is introduced to the result:

$$\text{ATD} = 143;\ \text{voltage} = 2.8039; \quad \text{ATD} = 142;\ \text{voltage} = 2.7843.$$

- Using linear interpolation: $143 - \dfrac{2.8039 - 2.8}{2.8039 - 2.7843} = 143 - \dfrac{0.0039}{0.0196} = 142.801$.

- Since the Sharp GP2Y0A02 IR sensor can measure any distance between 8 and 60 inches (20 cm ~ 150 cm), the calibrated data point should be between 50 and 90 points. More data points will provide a better estimation. With all the rounded-off *ATD* values, plot the curve $1/(R + k)$ versus *ATD*, with reference to the GP2D12 data.[2]

$$\frac{1}{R+k} = m(ATD) + b \quad \text{is in parallel with } y = mx + b$$

where R = distance range value ; m = slope of the function
k value is in the neighborhood of 1.
b = the projected line intersected with the vertical axis.

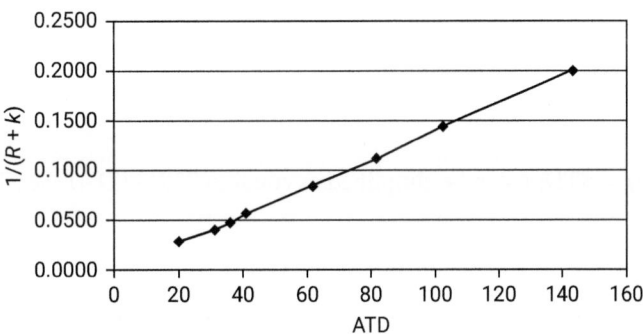

FIGURE 2.2: Round-off ATD versus 1/(R + k)

---

2 http://www.acroname.com/robotics/info/articles/irlinear/irlinear.html#e4

$$\frac{1}{R+k} = m(ATD)+b \quad \Rightarrow \quad 1 = m(ATD)[R+k]+b[R+k]$$

$$1 = R[m(ATD)+b]+k[m(ATD)+b] \quad \Rightarrow \quad 1-k[m(ATD)+b] = R[m(ATD)+b]$$

$$R = \frac{1-k[m(ATD)+b]}{k[m(ATD)+b]} = \frac{1}{k[m(ATD)+b]} - 1 = \frac{1/(mk)}{(ATD)+(b/m)} - 1$$

- After simplifying the equation for the distance R, as shown above, we have to do without much number crunching with the microprocessor. The only way is to round off the terms $1/(mk)$ and b/m with a constant. Using the quantized ATD value to estimate the distance range value, compute the offset from the calibrated value, and square the error.

The coefficients are computed with round-off from the m file (MATLAB):

$$m = 0.00140;\ b = -0.0007;\ k = 0.98$$

$$1/mk = 728.863 \approx 729 \quad \text{and} \quad b/m = -0.0007/0.0014 = -0.5.$$

TABLE 2.2A: Linearized ATD value versus round-off ATD value

Calibrated			Computed from ATD with round-off			
R (in cm)	Voltage (in V)	ATD (linearized)	ATD round-off	$1/mk$ round-off	b/m	R (in cm)
4	2.8	142.80	143	729	−0.5	4.1158
6	2	102.00	102	729	−0.5	6.1823
8	1.6	81.60	82	729	−0.5	7.9448
11	1.2	61.20	61	729	−0.5	11.0496
17	0.8	40.80	41	729	−0.5	17.0000
20	0.7	35.70	36	729	−0.5	19.5352
24	0.6	30.60	31	729	−0.5	22.9016
34	0.4	20.40	20	729	−0.5	36.3846

- Round (ATD −0.5) = ATD; R can be simplified as equal to 729/ATD −1. See Table 2.2b.

TABLE 2.2B

Estimated R	Actual R	Error	Squared error
4.1158	4	0.1158	0.0134
6.1823	6	0.1823	0.0332
7.9448	8	−0.0552	0.0030
11.0496	11	0.0496	0.0025
17.0000	17	0.0000	0.0000
19.5352	20	−0.4648	0.2160
22.9016	24	−1.0984	1.2064
36.3846	34	2.3846	5.6864
Sum of squared error			7.1610

- The total squared error for all data points can be computed with the following equation:

$$\text{Total squared error} = \sum_{k=1}^{num} \left(R_{estimated} - R_{calibrated} \right)^2$$

$$\text{Mean squared error} = \frac{1}{num} \sum_{k=1}^{num} \left(R_{estimated} - R_{calibrated} \right)^2$$

where

num is the total number of data points.

The following *m* file for MATLAB is to search the optimal squared error for the linear function.

HOW TO MODIFY THE FOLLOWING MATLAB FILE WITH YOUR CALIBRATED DATA

Replace the declaration for *ATD* and *R* arrays with your calibrated data. Change the upper limit for the most inside *For* loop to the number of calibrated data points for your data set. First run, the menu will ask you: Do you want to see more solutions? Choose No. Then the *m* file will show the optimal solution in the output file named *mse.txt*. If you open *mse.txt*, you will see the value for the mean squared error. Run the *m* file again, and choose Yes when the menu asks you: Do you want to see more solutions? It will ask you to input the value for total squared error threshold. Type in a value that is slightly bigger than the minimum squared error you got in the first run. In doing so, you will see various sets of values that are close to the optimal solution.

```
% This program computes the mean squared error for all data points.
fid = fopen('mse.txt','w');
ATD = [143,102,82,61,41,36,31,20];          % calibrated ATD values
R = [4,6,8,11,17,20,24,34];                 % calibrated Range values
choice=menu('Do you want to see more solutions?','Yes','No');
num=8; % num = # of data points
if choice==1,
   margin=input('Mean Squared Error Threshold = ? ');
end
errsq=500;                    % initialize the initial value
for n=1:40
   kk(n)=0.8+n*0.01;          % range of k values
   for i=1:50
     b(i)=-0.0008+i*0.00001;  % range of b values
     for j=1:8
       m(j)=0.00134+j*0.00001; % range of m values
       err_sq(j)=0.0;
       for k=1:num             % # of data points = 8
```

```
        top=1/(m(j)*kk(n));
        bottom=ATD(k)+b(i)/m(j);
        RR(k)=round(top)/round(bottom)−1;
        err(k)=R(k)−RR(k);
        err_sq(j)=err(k)^2+err_sq(j);
      end
      err_sq(j)=err_sq(j)/num;        % compute the mean value
      if choice==1
        if err_sq(j) < margin
          fprintf(fid,'m=%9.7f, b=%9.7f, k = %9.7f, TTL_SE=%12.8f\n', ...
            m(j),b(i),kk(n),err_sq(j));
        end
      end
      if err_sq(j) < errsq
        errsq=err_sq(j);
        A=m(j); B=b(i); C=kk(n);
      end
    end
   end
  end
end
fprintf(fid,'************* Optimal Solution *********************\n');
fprintf(fid,'m=%10.7f, b=%10.7f, k=%10.7f, Mean_Sq_Error=%12.8f\n',...
A,B,C,errsq);
fclose(fid);
```

Note: Plot your calibrated data as shown in Figure 2.2. Derive the *m*, *b*, and *k* values from the plot. The *k* value should be around 1.0. Modify the range of values for *m*, *b*, and *k* in the above MATLAB *m* file.

TABLE 2.3: 8-bit ATD versus rebounded signal voltage with V_{RH} = 5 V (1 of 2)

ATD	Voltage	ATD	Voltage	ATD	Voltage	ATD	Voltage
255	5.0000	214	4.1961	173	3.3922	132	2.5882
254	4.9804	213	4.1765	172	3.3725	131	2.5686
253	4.9608	212	4.1569	171	3.3529	130	2.5490
252	4.9412	211	4.1373	170	3.3333	129	2.5294
251	4.9216	210	4.1176	169	3.3137	128	2.5098
250	4.0863	209	4.0980	168	3.2941	127	2.4902
249	4.9020	208	4.0784	167	3.2745	126	2.4706
248	4.9020	207	4.0588	166	3.2549	125	2.4510
247	4.9020	206	4.0392	165	3.2353	124	2.4314
246	4.9020	205	4.0196	164	3.2157	123	2.4118
245	4.8039	204	4.0000	163	3.1961	122	2.3922
244	4.7843	203	3.9804	162	3.1765	121	2.3725
243	4.7647	202	3.9608	161	3.1569	120	2.3529
242	4.7451	201	3.9412	160	3.1373	119	2.3333
241	4.7255	200	3.9216	159	3.1176	118	2.3137
240	4.7059	199	3.9020	158	3.0980	117	2.2941
239	4.6863	198	3.8824	157	3.0784	116	2.2745
238	4.6667	197	3.8627	156	3.0588	115	2.2549
237	4.6471	196	3.8431	155	3.0392	114	2.2353
236	4.6275	195	3.8235	154	3.0196	113	2.2157
235	4.6078	194	3.8039	153	3.0000	112	2.1961
234	4.5882	193	3.7843	152	2.9804	111	2.1765
233	4.5686	192	3.7647	151	2.9608	110	2.1569
232	4.5490	191	3.7451	150	2.9412	109	2.1373
231	4.5294	190	3.7255	149	2.9216	108	2.1176
230	4.5098	189	3.7059	148	2.9020	107	2.0980
229	4.4902	188	3.6863	147	2.8824	106	2.0784
228	4.4706	187	3.6667	146	2.8627	105	2.0588
227	4.4510	186	3.6471	145	2.8431	104	2.0392
226	4.4314	185	3.6275	144	2.8235	103	2.0196
225	4.4118	184	3.6078	***143***	***2.8039***	***102***	***2.0000***
224	4.3922	183	3.5882	***142***	***2.7843***	101	1.9804
223	4.3725	182	3.5686	141	2.7647	100	1.9608
222	4.3529	181	3.5490	140	2.7451	99	1.9412
221	4.3333	180	3.5294	139	2.7255	98	1.9216
220	4.3137	179	3.5098	138	2.7059	97	1.9020
219	4.2941	178	3.4902	137	2.6863	96	1.8824
218	4.2745	177	3.4706	136	2.6667	95	1.8627

(Continued)

ATD	Voltage	ATD	Voltage	ATD	Voltage	ATD	Voltage
217	4.2549	176	3.4510	135	2.6471	94	1.8431
216	4.2353	175	3.4314	134	2.6275	93	1.8235
215	4.2157	174	3.4118	133	2.6078	92	1.8039

TABLE 2.3: 8-bit ATD versus rebounded signal voltage with V_{RH} = 5 V (2 of 2)

ATD	Voltage	ATD	Voltage	ATD	Voltage
91	1.7843	50	0.9804	9	0.1765
90	1.7647	49	0.9608	8	0.1569
89	1.7451	48	0.9412	7	0.1373
88	1.7255	47	0.9216	6	0.1176
87	1.7059	46	0.9020	5	0.0980
86	1.6863	45	0.8824	4	0.0784
85	1.6667	44	0.8627	3	0.0588
84	1.6471	43	0.8431	2	0.0392
83	1.6275	42	0.8235	1	0.0196
82	*1.6078*	*41*	*0.8039*	0	0.0000
81	*1.5882*	40	0.7843		
80	1.5686	39	0.7647		
79	1.5490	38	0.7451		
78	1.5294	37	0.7255		
77	1.5098	*36*	*0.7059*		
76	1.4902	*35*	*0.6863*		
75	1.4706	34	0.6667		
74	1.4510	33	0.6471		
73	1.4314	32	0.6275		
72	1.4118	*31*	*0.6078*		
71	1.3922	*30*	*0.5882*		
70	1.3725	29	0.5686		
69	1.3529	28	0.5490		
68	1.3333	27	0.5294		
67	1.3137	26	0.5098		
66	1.2941	25	0.4902		
65	1.2745	24	0.4706		
64	1.2549	23	0.4510		
63	1.2353	22	0.4314		
62	*1.2157*	*21*	*0.4118*		
61	*1.1961*	*20*	*0.3922*		
60	1.1765	19	0.3725		
59	1.1569	18	0.3529		

(*Continued*)

58	1.1373	17	0.3333	
57	1.1176	16	0.3137	
56	1.0980	15	0.2941	
55	1.0784	14	0.2745	
54	1.0588	13	0.2549	
53	1.0392	12	0.2353	
52	1.0196	11	0.2157	
51	1.0000	10	0.1961	

Note: The entries in bold and italic are corresponded to the calibrated data in Figure 2.2.

TABLE 2.4: 10-bit ATD versus rebounded signal voltage with V_{RH} = 5 V (1 of 7)

ATD	Voltage	ATD	Voltage	ATD	Voltage	ATD	Voltage
1023	5.0000	982	4.7996	941	4.5992	900	4.3988
1022	4.9951	981	4.7947	940	4.5943	899	4.3939
1021	4.9902	980	4.7898	939	4.5894	898	4.3891
1020	4.9853	979	4.7849	938	4.5846	897	4.3842
1019	4.9804	978	4.7801	937	4.5797	896	4.3793
1018	4.9756	977	4.7752	936	4.5748	895	4.3744
1017	4.9707	976	4.7703	935	4.5699	894	4.3695
1016	4.9658	975	4.7654	934	4.5650	893	4.3646
1015	4.9609	974	4.7605	933	4.5601	892	4.3597
1014	4.9560	973	4.7556	932	4.5552	891	4.3548
1013	4.9511	972	4.7507	931	4.5503	890	4.3500
1012	4.9462	971	4.7458	930	4.5455	889	4.3451
1011	4.9413	970	4.7410	929	4.5406	888	4.3402
1010	4.9365	969	4.7361	928	4.5357	887	4.3353
1009	4.9316	968	4.7312	927	4.5308	886	4.3304
1008	4.9267	967	4.7263	926	4.5259	885	4.3255
1007	4.9218	966	4.7214	925	4.5210	884	4.3206
1006	4.9169	965	4.7165	924	4.5161	883	4.3157
1005	4.9120	964	4.7116	923	4.5112	882	4.3109
1004	4.9071	963	4.7067	922	4.5064	881	4.3060
1003	4.9022	962	4.7019	921	4.5015	880	4.3011
1002	4.8974	961	4.6970	920	4.4966	879	4.2962
1001	4.8925	960	4.6921	919	4.4917	878	4.2913
1000	4.8876	959	4.6872	918	4.4868	877	4.2864
999	4.8827	958	4.6823	917	4.4819	876	4.2815
998	4.8778	957	4.6774	916	4.4770	875	4.2766
997	4.8729	956	4.6725	915	4.4721	874	4.2717
996	4.8680	955	4.6676	914	4.4673	873	4.2669
995	4.8631	954	4.6628	913	4.4624	872	4.2620

(*Continued*)

ATD	Voltage	ATD	Voltage	ATD	Voltage	ATD	Voltage
994	4.8583	953	4.6579	912	4.4575	871	4.2571
993	4.8534	952	4.6530	911	4.4526	870	4.2522
992	4.8485	951	4.6481	910	4.4477	869	4.2473
991	4.8436	950	4.6432	909	4.4428	868	4.2424
990	4.8387	949	4.6383	908	4.4379	867	4.2375
989	4.8338	948	4.6334	907	4.4330	866	4.2326
988	4.8289	947	4.6285	906	4.4282	865	4.2278
987	4.8240	946	4.6237	905	4.4233	864	4.2229
986	4.8192	945	4.6188	904	4.4184	863	4.2180
985	4.8143	944	4.6139	903	4.4135	862	4.2131
984	4.8094	943	4.6090	902	4.4086	861	4.2082
983	4.8045	942	4.6041	901	4.4037	860	4.2033

TABLE 2.4: 10-bit ATD versus rebounded signal voltage with V_{RH} = 5 V (2 of 7)

ATD	Voltage	ATD	Voltage	ATD	Voltage	ATD	Voltage
859	4.1984	818	3.9980	777	3.7977	736	3.5973
858	4.1935	817	3.9932	776	3.7928	735	3.5924
857	4.1887	816	3.9883	775	3.7879	734	3.5875
856	4.1838	815	3.9834	774	3.7830	733	3.5826
855	4.1789	814	3.9785	773	3.7781	732	3.5777
854	4.1740	813	3.9736	772	3.7732	731	3.5728
853	4.1691	812	3.9687	771	3.7683	730	3.5679
852	4.1642	811	3.9638	770	3.7634	729	3.5630
851	4.1593	810	3.9589	769	3.7586	728	3.5582
850	4.1544	809	3.9541	768	3.7537	727	3.5533
849	4.1496	808	3.9492	767	3.7488	726	3.5484
848	4.1447	807	3.9443	766	3.7439	725	3.5435
847	4.1398	806	3.9394	765	3.7390	724	3.5386
846	4.1349	805	3.9345	764	3.7341	723	3.5337
845	4.1300	804	3.9296	763	3.7292	722	3.5288
844	4.1251	803	3.9247	762	3.7243	721	3.5239
843	4.1202	802	3.9198	761	3.7195	720	3.5191
842	4.1153	801	3.9150	760	3.7146	719	3.5142
841	4.1105	800	3.9101	759	3.7097	718	3.5093
840	4.1056	799	3.9052	758	3.7048	717	3.5044
839	4.1007	798	3.9003	757	3.6999	716	3.4995
838	4.0958	797	3.8954	756	3.6950	715	3.4946
837	4.0909	796	3.8905	755	3.6901	714	3.4897
836	4.0860	795	3.8856	754	3.6852	713	3.4848

(*Continued*)

ATD	Voltage	ATD	Voltage	ATD	Voltage	ATD	Voltage
835	4.0811	794	3.8807	753	3.6804	712	3.4800
834	4.0762	793	3.8759	752	3.6755	711	3.4751
833	4.0714	792	3.8710	751	3.6706	710	3.4702
832	4.0665	791	3.8661	750	3.6657	709	3.4653
831	4.0616	790	3.8612	749	3.6608	708	3.4604
830	4.0567	789	3.8563	748	3.6559	707	3.4555
829	4.0518	788	3.8514	747	3.6510	706	3.4506
828	4.0469	787	3.8465	746	3.6461	705	3.4457
827	4.0420	786	3.8416	745	3.6413	704	3.4409
826	4.0371	785	3.8368	744	3.6364	703	3.4360
825	4.0323	784	3.8319	743	3.6315	702	3.4311
824	4.0274	783	3.8270	742	3.6266	701	3.4262
823	4.0225	782	3.8221	741	3.6217	700	3.4213
822	4.0176	781	3.8172	740	3.6168	699	3.4164
821	4.0127	780	3.8123	739	3.6119	698	3.4115
820	4.0078	779	3.8074	738	3.6070	697	3.4066
819	4.0029	778	3.8025	737	3.6022	696	3.4018

TABLE 2.4: 10-bit ATD versus rebounded signal voltage with V_{RH} = 5 V (3 of 7)

ATD	Voltage	ATD	Voltage	ATD	Voltage	ATD	Voltage
695	3.3969	654	3.1965	613	2.9961	572	2.7957
694	3.3920	653	3.1916	612	2.9912	571	2.7908
693	3.3871	652	3.1867	611	2.9863	570	2.7859
692	3.3822	651	3.1818	610	2.9814	569	2.7810
691	3.3773	650	3.1769	609	2.9765	568	2.7761
690	3.3724	649	3.1720	608	2.9717	567	2.7713
689	3.3675	648	3.1672	607	2.9668	566	2.7664
688	3.3627	647	3.1623	606	2.9619	565	2.7615
687	3.3578	646	3.1574	605	2.9570	564	2.7566
686	3.3529	645	3.1525	604	2.9521	563	2.7517
685	3.3480	644	3.1476	603	2.9472	562	2.7468
684	3.3431	643	3.1427	602	2.9423	561	2.7419
683	3.3382	642	3.1378	601	2.9374	560	2.7370
682	3.3333	641	3.1329	600	2.9326	559	2.7322
681	3.3284	640	3.1281	599	2.9277	558	2.7273
680	3.3236	639	3.1232	598	2.9228	557	2.7224
679	3.3187	638	3.1183	597	2.9179	556	2.7175
678	3.3138	637	3.1134	596	2.9130	555	2.7126
677	3.3089	636	3.1085	595	2.9081	554	2.7077

(*Continued*)

ATD	Voltage	ATD	Voltage	ATD	Voltage	ATD	Voltage
676	3.3040	635	3.1036	594	2.9032	553	2.7028
675	3.2991	634	3.0987	593	2.8983	552	2.6979
674	3.2942	633	3.0938	592	2.8935	551	2.6931
673	3.2893	632	3.0890	591	2.8886	550	2.6882
672	3.2845	631	3.0841	590	2.8837	549	2.6833
671	3.2796	630	3.0792	589	2.8788	548	2.6784
670	3.2747	629	3.0743	588	2.8739	547	2.6735
669	3.2698	628	3.0694	587	2.8690	546	2.6686
668	3.2649	627	3.0645	586	2.8641	545	2.6637
667	3.2600	626	3.0596	585	2.8592	544	2.6588
666	3.2551	625	3.0547	584	2.8543	543	2.6540
665	3.2502	624	3.0499	583	2.8495	542	2.6491
664	3.2454	623	3.0450	582	2.8446	541	2.6442
663	3.2405	622	3.0401	581	2.8397	540	2.6393
662	3.2356	621	3.0352	580	2.8348	539	2.6344
661	3.2307	620	3.0303	579	2.8299	538	2.6295
660	3.2258	619	3.0254	578	2.8250	537	2.6246
659	3.2209	618	3.0205	577	2.8201	536	2.6197
658	3.2160	617	3.0156	576	2.8152	535	2.6149
657	3.2111	616	3.0108	575	2.8104	534	2.6100
656	3.2063	615	3.0059	574	2.8055	533	2.6051
655	3.2014	614	3.0010	573	2.8006	532	2.6002

TABLE 2.4: 10-bit ATD versus rebounded signal voltage with V_{RH} = 5 V (4 of 7)

ATD	Voltage	ATD	Voltage	ATD	Voltage	ATD	Voltage
531	2.5953	490	2.3949	449	2.1945	408	1.9941
530	2.5904	489	2.3900	448	2.1896	407	1.9892
529	2.5855	488	2.3851	447	2.1848	406	1.9844
528	2.5806	487	2.3803	446	2.1799	405	1.9795
527	2.5758	486	2.3754	445	2.1750	404	1.9746
526	2.5709	485	2.3705	444	2.1701	403	1.9697
525	2.5660	484	2.3656	443	2.1652	402	1.9648
524	2.5611	483	2.3607	442	2.1603	401	1.9599
523	2.5562	482	2.3558	441	2.1554	400	1.9550
522	2.5513	481	2.3509	440	2.1505	399	1.9501
521	2.5464	480	2.3460	439	2.1457	398	1.9453
520	2.5415	479	2.3412	438	2.1408	397	1.9404
519	2.5367	478	2.3363	437	2.1359	396	1.9355
518	2.5318	477	2.3314	436	2.1310	395	1.9306
517	2.5269	476	2.3265	435	2.1261	394	1.9257

(Continued)

ATD	Voltage	ATD	Voltage	ATD	Voltage	ATD	Voltage
516	2.5220	475	2.3216	434	2.1212	393	1.9208
515	2.5171	474	2.3167	433	2.1163	392	1.9159
514	2.5122	473	2.3118	432	2.1114	391	1.9110
513	2.5073	472	2.3069	431	2.1065	390	1.9062
512	2.5024	471	2.3021	430	2.1017	389	1.9013
511	2.4976	470	2.2972	429	2.0968	388	1.8964
510	2.4927	469	2.2923	428	2.0919	387	1.8915
509	2.4878	468	2.2874	427	2.0870	386	1.8866
508	2.4829	467	2.2825	426	2.0821	385	1.8817
507	2.4780	466	2.2776	425	2.0772	384	1.8768
506	2.4731	465	2.2727	424	2.0723	383	1.8719
505	2.4682	464	2.2678	423	2.0674	382	1.8671
504	2.4633	463	2.2630	422	2.0626	381	1.8622
503	2.4585	462	2.2581	421	2.0577	380	1.8573
502	2.4536	461	2.2532	420	2.0528	379	1.8524
501	2.4487	460	2.2483	419	2.0479	378	1.8475
500	2.4438	459	2.2434	418	2.0430	377	1.8426
499	2.4389	458	2.2385	417	2.0381	376	1.8377
498	2.4340	457	2.2336	416	2.0332	375	1.8328
497	2.4291	456	2.2287	415	2.0283	374	1.8280
496	2.4242	455	2.2239	414	2.0235	373	1.8231
495	2.4194	454	2.2190	413	2.0186	372	1.8182
494	2.4145	453	2.2141	412	2.0137	371	1.8133
493	2.4096	452	2.2092	411	2.0088	370	1.8084
492	2.4047	451	2.2043	410	2.0039	369	1.8035
491	2.3998	450	2.1994	409	1.9990	368	1.7986

TABLE 2.4: 10-bit ATD versus rebounded signal voltage with V_{RH} = 5 V (5 of 7)

ATD	Voltage	ATD	Voltage	ATD	Voltage	ATD	Voltage
367	1.7937	326	1.5934	285	1.3930	244	1.1926
366	1.7889	325	1.5885	284	1.3881	243	1.1877
365	1.7840	324	1.5836	283	1.3832	242	1.1828
364	1.7791	323	1.5787	282	1.3783	241	1.1779
363	1.7742	322	1.5738	281	1.3734	240	1.1730
362	1.7693	321	1.5689	280	1.3685	239	1.1681
361	1.7644	320	1.5640	279	1.3636	238	1.1632
360	1.7595	319	1.5591	278	1.3587	237	1.1584
359	1.7546	318	1.5543	277	1.3539	236	1.1535
358	1.7498	317	1.5494	276	1.3490	235	1.1486

(*Continued*)

ATD	Voltage	ATD	Voltage	ATD	Voltage	ATD	Voltage
357	1.7449	316	1.5445	275	1.3441	234	1.1437
356	1.7400	315	1.5396	274	1.3392	233	1.1388
355	1.7351	314	1.5347	273	1.3343	232	1.1339
354	1.7302	313	1.5298	272	1.3294	231	1.1290
353	1.7253	312	1.5249	271	1.3245	230	1.1241
352	1.7204	311	1.5200	270	1.3196	229	1.1193
351	1.7155	310	1.5152	269	1.3148	228	1.1144
350	1.7107	309	1.5103	268	1.3099	227	1.1095
349	1.7058	308	1.5054	267	1.3050	226	1.1046
348	1.7009	307	1.5005	266	1.3001	225	1.0997
347	1.6960	306	1.4956	265	1.2952	224	1.0948
346	1.6911	305	1.4907	264	1.2903	223	1.0899
345	1.6862	304	1.4858	263	1.2854	222	1.0850
344	1.6813	303	1.4809	262	1.2805	221	1.0802
343	1.6764	302	1.4761	261	1.2757	220	1.0753
342	1.6716	301	1.4712	260	1.2708	219	1.0704
341	1.6667	300	1.4663	259	1.2659	218	1.0655
340	1.6618	299	1.4614	258	1.2610	217	1.0606
339	1.6569	298	1.4565	257	1.2561	216	1.0557
338	1.6520	297	1.4516	256	1.2512	215	1.0508
337	1.6471	296	1.4467	255	1.2463	214	1.0459
336	1.6422	295	1.4418	254	1.2414	213	1.0411
335	1.6373	294	1.4370	253	1.2366	212	1.0362
334	1.6325	293	1.4321	252	1.2317	211	1.0313
333	1.6276	292	1.4272	251	1.2268	210	1.0264
332	1.6227	291	1.4223	250	1.2219	209	1.0215
331	1.6178	290	1.4174	249	1.2170	208	1.0166
330	1.6129	289	1.4125	248	1.2121	207	1.0117
329	1.6080	288	1.4076	247	1.2072	206	1.0068
328	1.6031	287	1.4027	246	1.2023	205	1.0020
327	1.5982	286	1.3978	245	1.1975	204	0.9971

TABLE 2.4: 10-bit ATD versus rebounded signal voltage with V_{RH} = 5 V (6 of 7)

ATD	Voltage	ATD	Voltage	ATD	Voltage	ATD	Voltage
203	0.9922	162	0.7918	121	0.5914	80	0.3910
202	0.9873	161	0.7869	120	0.5865	79	0.3861
201	0.9824	160	0.7820	119	0.5816	78	0.3812
200	0.9775	159	0.7771	118	0.5767	77	0.3763
199	0.9726	158	0.7722	117	0.5718	76	0.3715

(*Continued*)

ATD	Voltage	ATD	Voltage	ATD	Voltage	ATD	Voltage
198	0.9677	157	0.7674	116	0.5670	75	0.3666
197	0.9629	156	0.7625	115	0.5621	74	0.3617
196	0.9580	155	0.7576	114	0.5572	73	0.3568
195	0.9531	154	0.7527	113	0.5523	72	0.3519
194	0.9482	153	0.7478	112	0.5474	71	0.3470
193	0.9433	152	0.7429	111	0.5425	70	0.3421
192	0.9384	151	0.7380	110	0.5376	69	0.3372
191	0.9335	150	0.7331	109	0.5327	68	0.3324
190	0.9286	149	0.7283	108	0.5279	67	0.3275
189	0.9238	148	0.7234	107	0.5230	66	0.3226
188	0.9189	147	0.7185	106	0.5181	65	0.3177
187	0.9140	146	0.7136	105	0.5132	64	0.3128
186	0.9091	145	0.7087	104	0.5083	63	0.3079
185	0.9042	144	0.7038	103	0.5034	62	0.3030
184	0.8993	143	0.6989	102	0.4985	61	0.2981
183	0.8944	142	0.6940	101	0.4936	60	0.2933
182	0.8895	141	0.6891	100	0.4888	59	0.2884
181	0.8847	140	0.6843	99	0.4839	58	0.2835
180	0.8798	139	0.6794	98	0.4790	57	0.2786
179	0.8749	138	0.6745	97	0.4741	56	0.2737
178	0.8700	137	0.6696	96	0.4692	55	0.2688
177	0.8651	136	0.6647	95	0.4643	54	0.2639
176	0.8602	135	0.6598	94	0.4594	53	0.2590
175	0.8553	134	0.6549	93	0.4545	52	0.2542
174	0.8504	133	0.6500	92	0.4497	51	0.2493
173	0.8456	132	0.6452	91	0.4448	50	0.2444
172	0.8407	131	0.6403	90	0.4399	49	0.2395
171	0.8358	130	0.6354	89	0.4350	48	0.2346
170	0.8309	129	0.6305	88	0.4301	47	0.2297
169	0.8260	128	0.6256	87	0.4252	46	0.2248
168	0.8211	127	0.6207	86	0.4203	45	0.2199
167	0.8162	126	0.6158	85	0.4154	44	0.2151
166	0.8113	125	0.6109	84	0.4106	43	0.2102
165	0.8065	124	0.6061	83	0.4057	42	0.2053
164	0.8016	123	0.6012	82	0.4008	41	0.2004
163	0.7967	122	0.5963	81	0.3959	40	0.1955

TABLE 2.4: 10-bit ATD versus rebounded signal voltage with V_{RH} = 5 V (7 of 7)

ATD	Voltage
39	0.1906
38	0.1857
37	0.1808
36	0.1760
35	0.1711
34	0.1662
33	0.1613
32	0.1564
31	0.1515
30	0.1466
29	0.1417
28	0.1369
27	0.1320
26	0.1271
25	0.1222
24	0.1173
23	0.1124
22	0.1075
21	0.1026
20	0.0978
19	0.0929
18	0.0880
17	0.0831
16	0.0782
15	0.0733
14	0.0684
13	0.0635
12	0.0587
11	0.0538
10	0.0489
9	0.0440
8	0.0391
7	0.0342
6	0.0293
5	0.0244
4	0.0196
3	0.0147
2	0.0098
1	0.0049
0	0.0000

Note: If you are using 10-bit ATD, the coefficients (*m*, *b*, *k*) will be different from the coefficients for 8-bit ATD. For 8-bit ATD, you only need to read the high byte (8 bits) of the result register if the data is left justified, and if the data is right justified, you will need to read the low byte. As for the 10-bit ATD, you will have to read all two bytes of the result register. In general, 10-bit ATD will provide a more accurate result. Make sure you have a good set of calibrated data; more data points will be more accurate. If the set of calibrated data is accurate enough, the *m* file should be able to help you to solve for the optimal value for the coefficients (*m*, *b*, *k*).

PROJECT #3:
Measure the Rotations per Minute of a Brushless DC Motor

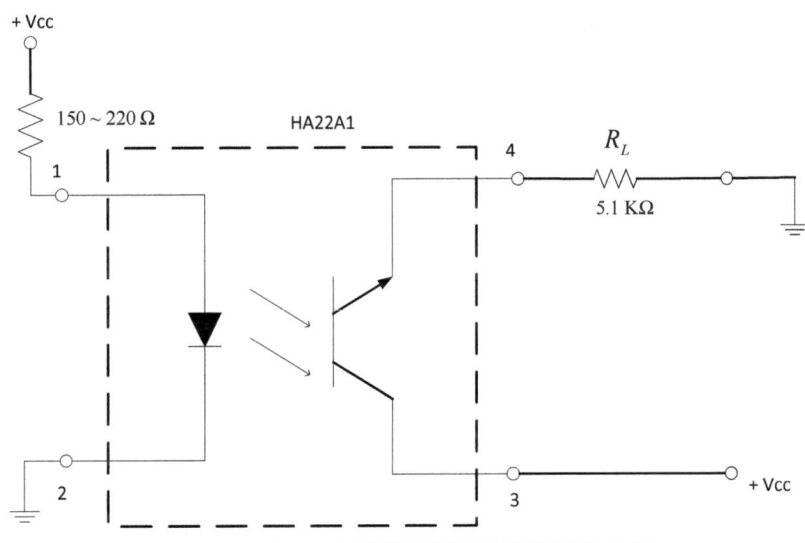

FIGURE 3.1: The circuit diagram for the slotted optical switch

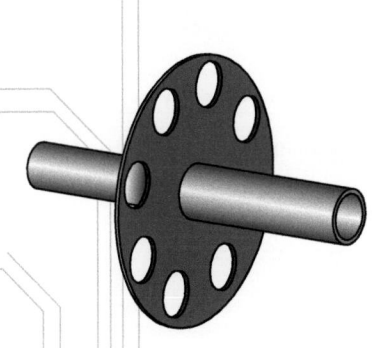

FIGURE 3.2: Optical encoder wheel

 Link for component: Optical Encoder Wheel:
Search 08DISK or Encoder Wheel in the following link:
http://www.elexp.com (Electronix Express)

The optical encoder wheel can be purchased from Electronix Express. This optical encoder wheel can fit a shaft between 0.08″ and 0.11″.

Most DC brushless motors have a shaft size of 0.08″. Using the heat shrink to seal the DC motor shaft (one or two layers), and then forcing the optical encoder wheel onto the DC motor shaft will enable the DC motor shaft to fit the optical encoder wheel tightly. The experimental setup is shown in Figure 3.3.

There are eight holes in the wheel encoder. Use input capture to capture either the rising edge or the falling edge. The time interval between the first rising (falling) edge and the ninth rising (falling) edge will provide the period for one rotation. You can also take the average for either four periods or eight periods. Beware of the timer overflow flag (TOF).

Note: Shifting the binary value to the right by two bits is equivalent to dividing the value by four. Shifting the binary value to the right by three bits is equivalent to dividing the value by eight.

Your setup should continuously monitor the rotations per minute (RPM) of your DC motor, and display the RPM on the LCD. The RPM display should be rounded up in integer. First you have to determine how many digits will be required to displace the RPM value. If there is a jump in the RPM value when the supplied voltage is kept constant, that means you have to pay attention to the TOF. Please review the Timer Overflow Count spreadsheet program on my webpage. The RPM of a DC brushless motor is directly proportional to the supplied DC voltage for the DC motor.

Use Microsoft Excel to plot the RPM versus the voltage supplied for the DC brushless motor and add it to your report. The supplied voltage should be between 3~10 volts.

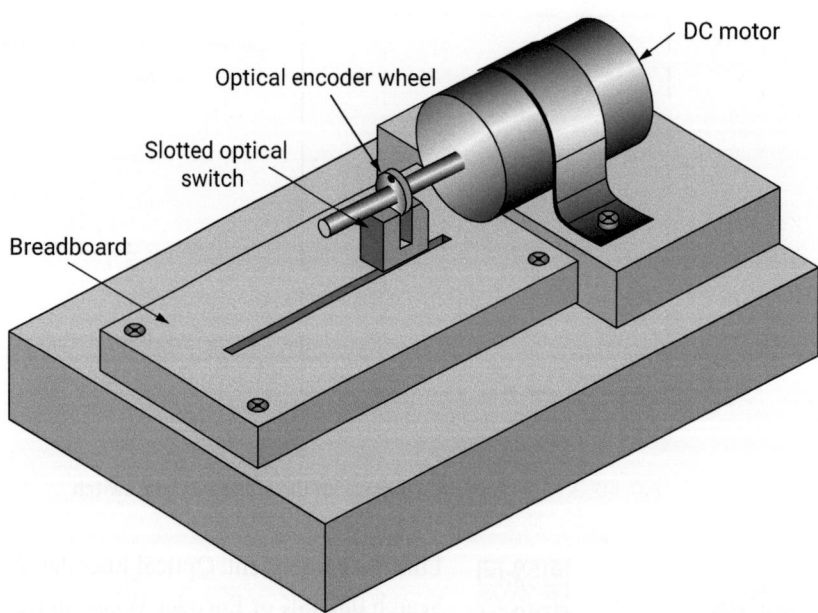

FIGURE 3.3: Setup for RPM measurement of a brushless DC motor

PROJECT #4:
LCD Timer/Stopwatch

Use a 2×20 or 2×24 character LCD for display and a timer capable of handling any time interval within twenty-four hours.

With the timer function, select a method to count 1 second. The display should have two digits for the hour and two digits for the minute. You will have to determine the total number of E cycles (overhead plus the delay counts) for the delayed routine such that it is providing an accurate count for 1 second. The accuracy plays a major role in this design.

```
delay10ms    movb    #$90,TSCR1         ; enable TCNT and fast flags clear
             movb    #$06,TSCR2         ; configure prescale factor to 64
             movb    #$01,TIOS          ; enable OC0
             ldd     TCNT
             addd    #3750              ; start an output compare operation
             std     TC0                ; with 10 ms time delay
             brclr   TFLG1,$01,*        ; if equal, C0F in TFLG1 is set to 1
             rts
```

The first row will illustrate the digits, as shown in Table 4.1 below:

TABLE 4.1

Time	0	0	:	1	0	:	3	0
Variables	h1	h0		m1	m0		s1	s0
Unit	Hour			Minute			Second	

Note: The labels for hour, minute, and second are not required for the LCD display.

In general, the commercial LCD timer does not have a display for seconds, but for verification purposes, it is better to keep the display for seconds so that you can compare your timer with the timer from your cellular phone (smartphones usually display the timer in minutes, seconds, and hundredths of a second). Commercial LCD timers usually do not have two digits for the hour—most of them only provide a single digit display for the hour. Since this project is the initial step for designing an LCD digital clock, it will be easier to keep the digital clock format for the time being.

You can use the two switches (S_1 and S_2) on Dragon12-JR for the following functions:

- Use S_1 on the Dragon12-JR to shift the LCD cursor to the left.

- Use S_2 is to increment the value for the digit marked with the cursor.

- Range of value: Least significant digit (s0) for second display: 0 ~ 9
  Most significant digit (s1) for second display: 0 ~ 5
  Least significant digit (m0) for minute display: 0 ~ 9
  Most significant digit (m1) for minute display: 0 ~ 5
  Least significant digit (h0) for hour display: 0 ~ 9
  Most significant digit (h1) for hour display: 0 ~ 2

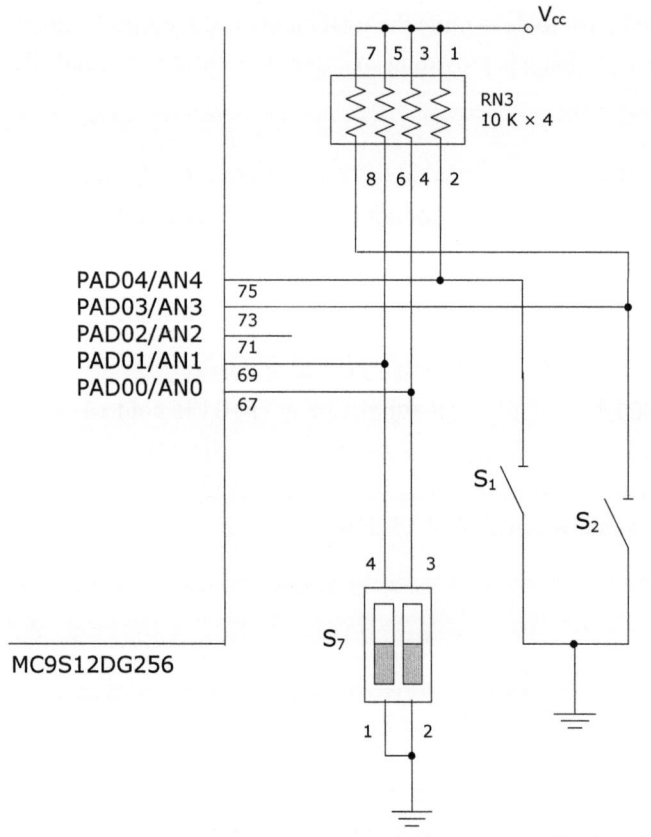

Note: S_1 and S_2 are both push-button switches. When it is pressed, it makes contact with the ground (digital input = 0); as soon as it is released, the switch will be back to open position (digital input = 1).

FIGURE 4.1: The circuit configuration for switches S_1 and S_2 in Dragon12-JR

- There is no carrier for the next digit, the increment of each digit is for itself only, and it is done in ASCII code; e.g., 0 = $30, 5 = $35, 9 = $39, : = $3A.

- After the cursor slides through all six digits for the time display, the timer will start. If you make a mistake in setting one of the digits, you have to start the time setting again.

- When the timer reaches the last 5 minutes, it should give a short buzzer for about 3 seconds.

- When the timer counts down to zero, the timer should sound the buzzer for 30 seconds. Then the timer will keep the previous preset time on display after the 30-second buzzer.

- On Dragon12-JR, switch S_1 is tied to PAD04 and switch S_2 is tied to PAD03. These two pins are enabled for digital inputs and they are either ON or OFF.

```
      movb    $18,ATD0DIEN            ; set digital input for PAD03 & PAD04
```

B7	B6	B5	B4	B3	B2	B1	B0	ATD0DIEN
0	0	0	1	1	0	0	0	$18

Use the timer counter reset enable (TCRE) bit in TSCR2

```
      set_TCRE    movb    #$80,TSCR1    ; configure TCNT for 200 ms.
                  movb    #$0F,TSCR2    ; TCRE = 1, prescaler = 128
                  movb    #$80,TIOS     ; enable output compare for Ch. 7
                  movw    #37500,TC7
                  rts
```

$$\text{The number of count} = \frac{(\text{delay time}) \times \text{MCLK}}{\text{Prescale Factor}} = \frac{(200 \times 10^{-3})(24 \times 10^6)}{128} = 37,500$$

With TCRE set at 1, when the timer counter, TCNT reaches the value 37,500 at TC7, the timer counter will reset to 0. That will be equivalent to a 200 ms delay. After running the 200 ms delay five times, it will be equivalent to 1 second. One precaution—while you are enabling TCRE, the TCNT might be greater than 37,500; if so, then you have to reset it to 0 by adding a number to overflow the TCNT.

Note: If you are using free running timer, then you have to watch out for the timer overflow flag, and carry out an extra step to determine the exact counts.

- With the time display starting at DDRAM address $02, the instruction is shown below:

```
      cursor  db      $8B                 ; cursor at DDRAM address $0B
              :
              :
              ldaa    #$82                ; specify DDRAM address ($02)
              jsr     cmd2LCD
              ldx     #msg1               ; time display 00:00:00
```

00	01	**0**	**0**	:	**0**	**0**	:	**0**	**0**	0A	0B	0C	0D	0E	0F	10	11	12	13
40	41	42	43	44	45	46	47	48	49	4A	4B	4C	4D	4E	4F	50	51	52	53

FIGURE 4.2: Time display format on LCD

PROS AND CONS OF ASCII CODE AND THE DECIMAL INCREMENT DISPLAY

- Timer: the user has to set the time and let the timer count down to 0 and sound the buzzer alarm.

- Stopwatch: the clock will start from 0 and terminate from a push-button switch or triggering mechanism. It runs the same routine as a military time or a regular time clock.

- If the user decides to use S_1 to increment the hours, and S_2 to increment the minutes, they have to press S_2 thirty times for 30 minutes, or forty times for 40 minutes, and there won't be any setting for seconds. The lifetime of those push-button switches is up to a finite number of presses (e.g., 10,000 times).

- The increment will follow the conventional format: 10 to 11, 19 to 20, 25 to 26, and etc. Each value will be decoded into ASCII code for the most significant digit and for the least significant digit before they can be displayed on the LCD.

- Incrementing each digit with the ASCII code can be more complex, but once the algorithm is established, the display algorithm is quite simple.

- Special attention has to be given in decrementing ASCII code for 0.

    For s0: $30 - 1 = $2F $2F + $0A = $39 $39 ⇨ digit 9

    For s1: $30 - 1 = $2F $2F + $06 = $35 $35 ⇨ digit 5

    Note: s0 stands for the least significant digit for the second display

            s1 stands for the most significant digit for the second display

    Time: 00:01:00   decrement by 1 second   00:00:59.

character	/	0	1	2	3	4	5	6	7	8	9	:
ASCII code	$2F	$30	$31	$32	$33	$34	$35	$36	$37	$38	$39	$3A

The following flowchart is for the two digits in the second display, and the decrement is using ASCII code. The same approach can be utilized for the two digits in the minute display.

Time: 01:00:00   decrement by 1 second   00:59:59

There are two digits in the hour display, but the constraints for this LCD timer are for any time interval less than or equal to 24 hours.

Time: 24:00:00   decrement by 1 second   23:59:59

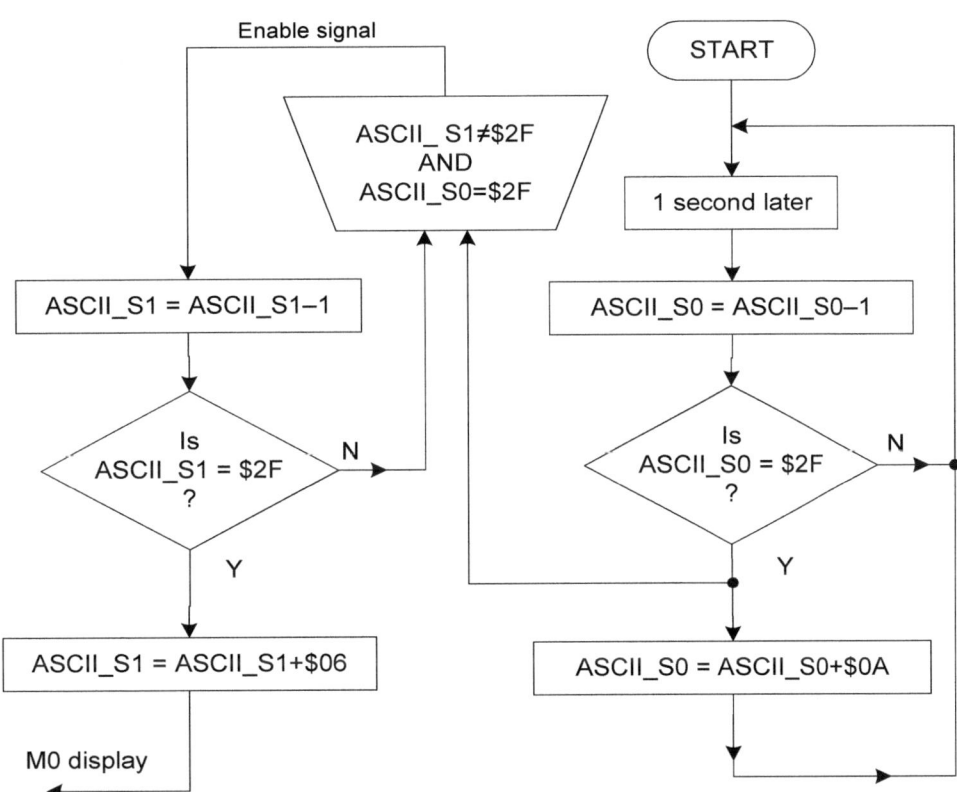

FIGURE 4.3: Flowchart for the second display in ASCII

You can compare your timer with the timer in your smartphone. The accuracy of your timer should be no more than ± 1 second within 15 minutes.

HOW TO CONVERT THIS TIMER TO A STOPWATCH

For a stopwatch, there is no initial time set; it always starts from zero. You can use switch S_1 to start the stopwatch, and switch S_2 to show the lap time. Press S_1 again to stop the stopwatch. The maximum number of laps is four. After pressing S_1 to stop the stopwatch, the LCD will display lap time #1 and lap time #2 in the first row, lap time #3 and lap time #4 in the second row. Pressing S_2 again will clear the display on the LCD.

PROJECT #5:
IR Triggering Control for Time Elapse

Sharp GP2D15 is a triggered infrared (IR) sensor. This sensor will generate a triggering pulse when it senses an object within 9.5 inches of the sensor. The slider setup is as shown in Figure 5.1.

The slider has two GP2D15 sensors: one is mounted at the top and the other one is at the bottom of the slider. The top one is to trigger the stopwatch to start, and the bottom one is to trigger the stopwatch to stop. Of course, the digital stopwatch from Project 4 will be utilized to display the time lapse for the object to roll down the slider on the LCD. You can use a ball or a can to roll down the slider. How to design the triggering component is the key for this project, not the speed.

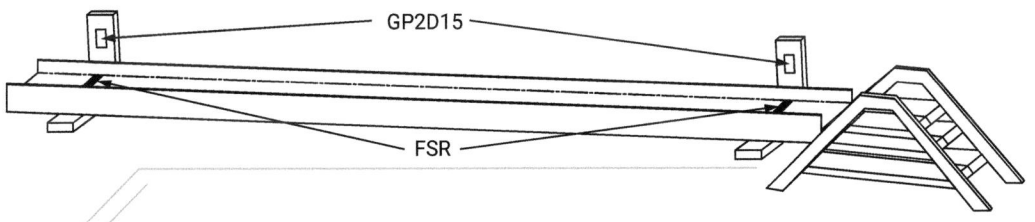

FIGURE 5.1: Setup for the slider

Note: The bracket for mounting the infrared (IR) sensor should be able to slide out or in from the slider. The distance between the two IR sensors is about four feet. The cables connecting the IR sensor to the microprocessor development board should be more than two feet in length. The two pieces of tape-like materials next to the IR sensors are force sensitive resistors (FSR), which is for another triggering control format.

ANOTHER TRIGGERING CONTROL METHOD WITH FSR

The force sensitive resistor (FSR) provides a relatively low resistance when it is pressed; and the stronger the pressing force is, the lower the output resistance will become.[1] When no pressure is being applied on the FSR, its resistance is greater than 1 MΩ. The FSR can sense any applied force in the range of 0.1 to 10 Kg. Implement a voltage divider with an FSR and a fixed resistor, such as 10 KΩ. If the supplied voltage to the voltage divider is 5 V, when the FSR is not pressed, the majority of the voltage will drop across the FSR. If the FSR is pressed hard, then majority of the voltage will drop across the 10 KΩ resistor. The rising edge of the 10 KΩ resistor voltage can be used to trigger the timer to start or stop. The rolling object has to be selected with various constraints; it has to be heavy enough to put pressure on the FSR and roll down the slope. The object should be unbreakable, and a catcher is recommended to stop the rolling object.

---

1   *https://www.sparkfun.com/products/9674*

PROJECT #6:
LCD Digital Clock (Military Time) with Alarm

Use a 2×16 or 2×20 or 2×24 character LCD for display.

This is a continuation of Project #4. Modify Project #4 to design a free-running digital clock for military time. If you got the stopwatch working, it will be quite easy to modify the program from Project #4.

The first row will illustrate the digits as shown below:

0	8	:	1	0	:	3	0
Hour			Minute			Second	
1	2	:	2	0	:	4	5
Hour			Minute			Second	
2	3	:	5	9	:	5	9
Hour			Minute			Second	

Note: The labels for hour, minute, and second are not required on the LCD display.

- Design a free-running digital clock for military time.
- Keep the same format as you had in Project #4 for the timer.
- For military time, 23:59:59 will be reset to 00:00:00 after 1 second.
- Use S1 to enable the alarm menu, and S2 for the main menu.
- While setting the time or the alarm, the clock is still running in the background.
- There is no indicator for AM or PM in military time.

The main program is separated into two routines: the main screen menu (S1:ALARM ; S2: TIME) and the reset routine (S1: LEFT : S2: INCR). The main screen routine begins after the initialization of the timer interrupt and the resetting of the LCD. The main screen menu displays the time as well as the menu options (S1:ALARM : S2:TIME).

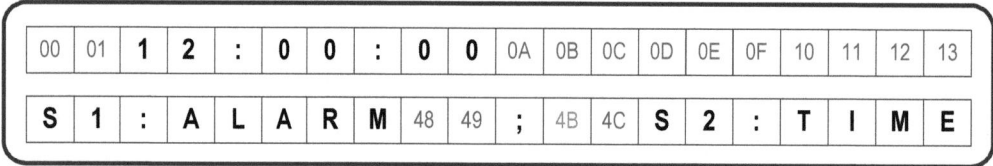

FIGURE 6.1: Main menu display on LCD

Unfortunately the character-based LCD cannot display the digit in subscript, the main screen menu S1 represents for switch S_1 and S2 represents for switch S_2. If switch S_1 is pressed, the reset routine is enabled, the alarm time can be set, and the cursor (underline) will indicate the location of the digit that can be incremented.

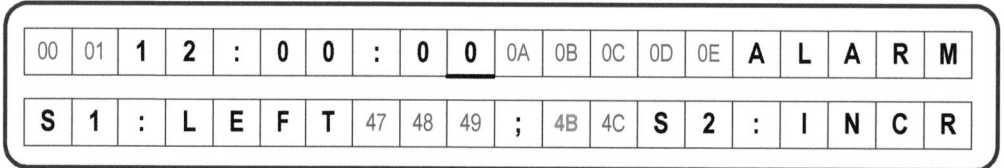

FIGURE 6.2: Alarm reset menu on LCD

If switch S_2 is pressed, the reset routine is enabled, the time can be set, and the cursor (underline) will indicate the location of the digit that can be incremented. Both reset routines are identical.

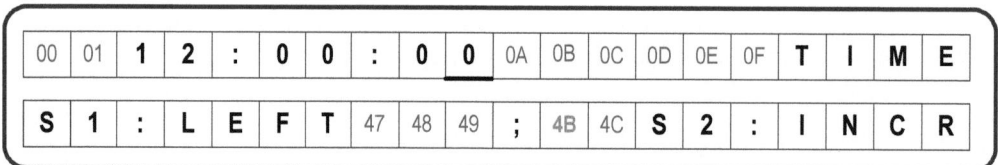

FIGURE 6.3: Time reset menu on LCD

By choosing S1 or S2 in the main menu, the operator can have the choice of setting the time or the alarm time. Both of these options use the reset routine, but when setting the time, the clock is halted in the background.

The reset routine enables the cursor and allows the user to set the digits for the time manually. In the reset mode, S1 allows the cursor to move left only and S2 to increment the underlined digit with the cursor, except the increment is limited to that digit only. A space constant is used to label the digit, which can be updated.

The digits are displayed on the top row of the LCD and the menu on the second or bottom line. After the reset routine is finished, the microprocessor will return to the main screen menu. The simplified logic of the program is shown below.

You can compare your designed clock with the clock in your cellular phone. The accuracy of your clock should not be more than ± 1 second within 30 minutes.

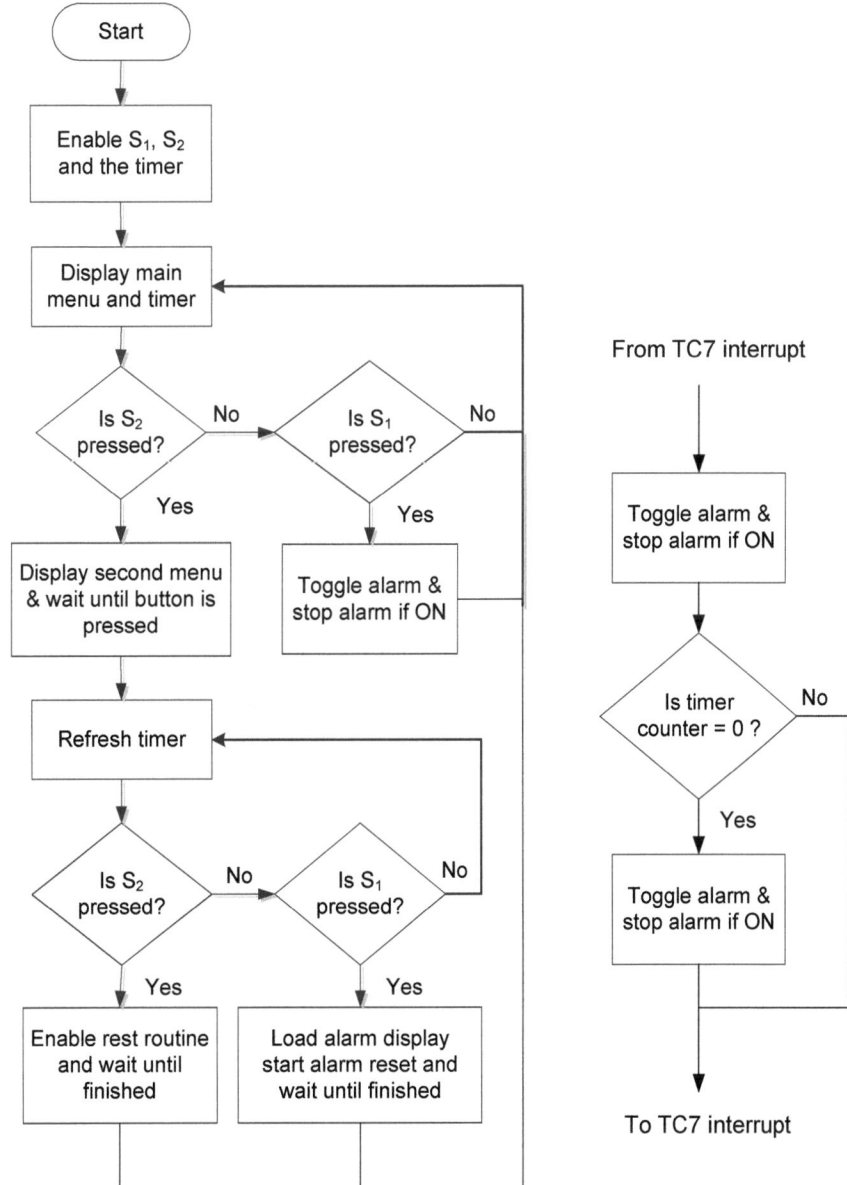

FIGURE 6.4: Flowchart for the LCD clock with alarm

PROJECT #7:
LCD Digital Clock with Alarm (AM/PM)

Use a 2×16 or 4×16 character LCD for display.

With the timer function, select a method to count 1 second. The display should have two digits for the hour, two digits for the minute, and two digits for the second. You will have to determine the total number of E cycles (overhead plus the delay counts) for the delayed routine such that it is providing an accurate count for 1 second. The accuracy plays a major role in this design.

The first row will illustrate the digits as shown below:

PM	0	0	:	1	0	:	3	0
	Hour			Minute			Second	

Note: The labels for hour, minute, and second are not required on the LCD display.

- Design a free-running digital clock for regular time with AM/PM display.

- Either one or both colon symbols should blink when the clock is running.

- Implement one of the switches on your Dragon12-JR to allow you to set the time. You can use the S_1 switch to set the hour and the S_2 switch to set the minute. This will let you to check whether the hour count will be incremented after 60 minutes. While you are setting the time or alarm, the clock is still running in the background.

- For normal time, AM 11:59:59 will increment to PM 12:00:00 in 1 second, and PM 11:59:59 will increment to AM 12:00:00 in 1 second.

- The difference in military time and regular time is in the LCD display. The logic of your program can still run military time in alarm mode, except when the time is 13:10:30 the LCD display should be PM 01:10:30.

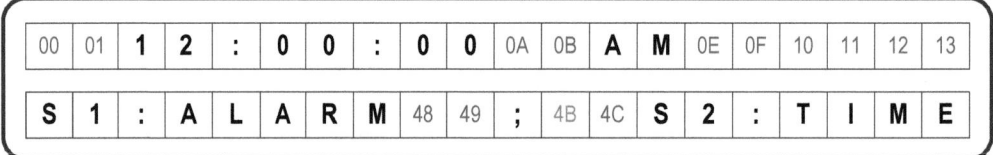

FIGURE 7.1: Main menu display on LCD

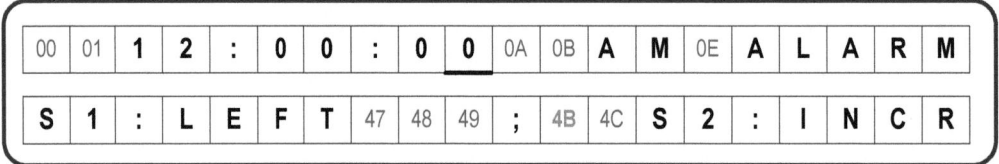

FIGURE 7.2: Alarm reset menu on LCD

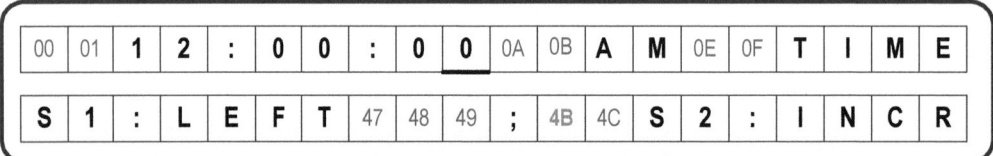

FIGURE 7.3: Time reset menu on LCD

The reset routine enables the cursor and allows the user to set the digits for the time manually. In the reset mode, S1 allows the cursor to move left only and S2 to increment the underlined digit with the cursor, except the increment is limited to that digit only. A space constant is used to label the digit that can be updated. After the cursor is moved past the hour place, the operator can choose AM or PM. The digits are displayed on the top row of the LCD and the menu on the second or bottom line. After the reset routine is finished, the microprocessor will return to the main screen menu.

You can compare your designed clock with the clock in your cellular phone.

The accuracy of your clock should not be more than ± 1 second within 30 minutes.

PROJECT #8:
Home Security System

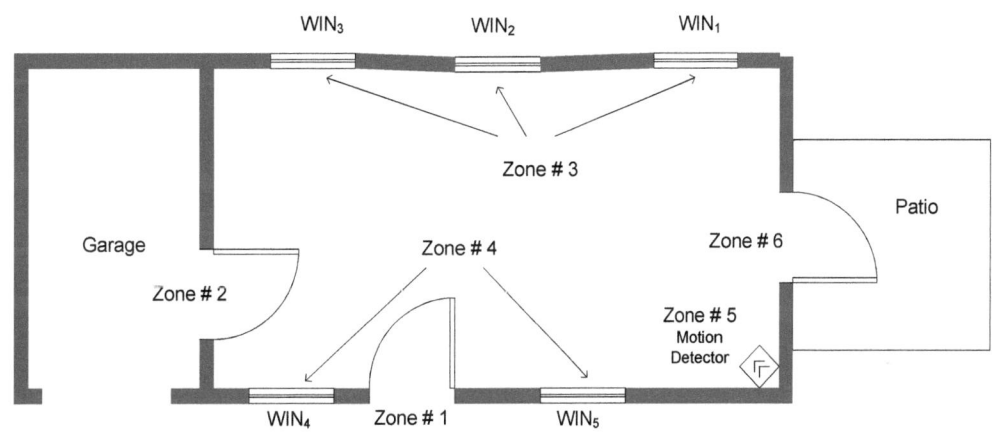

Zone #1:	Main entrance door	Zone #2:	Entrance from garage
Zone #3:	Windows #1, 2, and 3	Zone #4:	Windows #4 and 5
Zone #5:	Motion detector	Zone #6:	Patio door

FIGURE 8.1: A simple house layout

Only zones #1 and #2 (not zone #6) will allow the user to exit within 30 seconds after activating the alarm system.

Magnetic switches are used to monitor all the windows and doors. When the door or window is closed (opened), the switch is at the upper (lower) position, as shown in Figure 8.2.

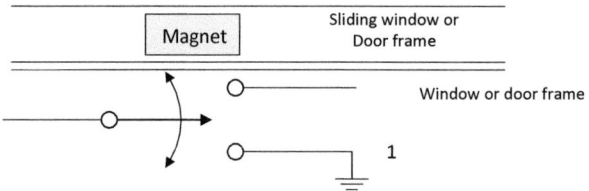

FIGURE 8.2:   A magnetic switch embedded inside a window or door frame

When the alarm system is activated, any break-in will activate the alarm. If a window or door in any zone is open, the alarm system cannot be activated unless that particular zone is by-passed. When the alarm is activated, and you are coming in from the outside, the alarm buzzer (first one) will come on as soon as you open the main door. This alarm buzzer (first one) will stay on until you type in your 4-digit PIN number. Within 30 seconds, if the 4-digit PIN number is not entered, then the high-pitch buzzer (second one) up in the attic will come on to alert your neighbors. That means the two buzzers will be on simultaneously if the 4-digit PIN number is not entered within 30 seconds after the entry. If there is a hesitation between two consecutive digits while you are entering the PIN, you have to enter the 4-digit PIN again; the PIN has to be entered without any hesitation. As soon as the PIN number is entered, both buzzers should turn off.

To bypass any particular zone, type the 4-digit PIN / Security Code + {6} key + zone number in single digit. Note the {6} key is the bypass key. When finished, the keypad will momentarily display a "Bypass" message for each bypassed zone number. Arm the system as usual. When armed, the arming message is displayed with "ZONE BYPASSED." Enter your security code and press the {6} BYPASS key to display the bypassed zones before arming. In case of a mistake, you can press the {*} key and the zone number to cancel the bypass zone. Even though the alarm system is not activated, there should be two or three beeps when someone opens any window or door, except for the motion detector. Store your 4-digit PIN onto the stack. After you activate the alarm system, you can only exit from zone #1 or #2 (not zone #6). Use {#} and {2} to activate the alarm system for Away mode, and use {#} and {3} to activate the alarm system for Stay mode.

This project covers keypad interface, the programmable timer in your HCS12, and of course the logic function of the security alarm system. Please remember to comment in your assembly statement. Provide a well-organized flow chart of your design in your report.

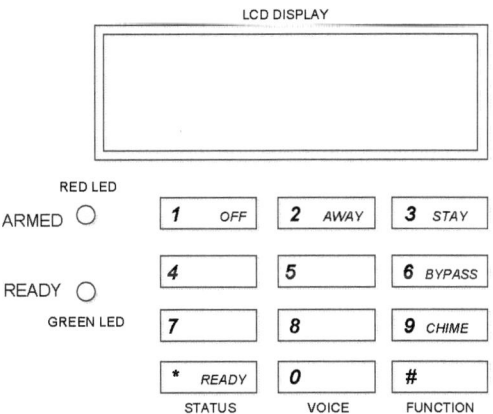

FIGURE 8.3: Home security system layout

- When entering codes and commands, sequential key depressions must be made within 2 seconds of one another. If there is any hesitation between entering keys, the entry will be aborted and must be repeated from the beginning.

- If there is a mistake while entering the PIN, stop, press the {*} key, and then start over. If you stop in the middle while entering a code, pause and let the system abort your entry, and then start the entry again.

USING THE {*} KEY

- Before arming your system, all protected doors, windows, and other protected zones must be closed or bypassed, otherwise a "Not Ready" message will be displayed on the LCD.

- By pressing the {*} key only, the system will display all faulted zones.

- Secure the displayed zones or bypass the displayed zones; the green LED will be ON when all protection zones have been either secured or bypassed. Then the system is ready to be activated.

ARMING COMMANDS WITH {2} OR {3} KEYS

Before arming the system, close all perimeter doors and windows and make sure the Ready to Arm message is displayed on the LCD (see Table 8.1).

TABLE 8.1

Mode	Press these keys…	Keypad confirms by …
Stay	Security Code + {3} key	Three beeps
		Armed STAY message displayed on LCD
		Red ARMED indicator is ON
		The motion detector is bypassed in Stay Mode.
Away	Security Code + {2} key	Two beeps, beeping for duration of exit delay
		Armed AWAY message displayed on LCD
		Red ARMED indicator is ON
		Leave the premises through an entry/exit door (only applies to zones 1 and 2) during the exit delay period to avoid causing an alarm. The keypad beeps in a faster rate during the last 10 seconds of the exit delay.

QUICK ARMING

The {#} key can be pressed as a replacement for the security code when arming the system in any of its arming modes (Stay or Away). The security code has to be entered when manually disarming the system.

GREEN/RED LED

Green light: Green LED is on when the system is disarmed and ready to be armed (no open zones). If the system is disarmed and the green light is OFF, it indicates the system is not ready (one or more zones are not secured).

Red light: Red LED is on when system is armed in AWAY or STAY mode and exit delay has expired.
Slow flashing = system armed STAY and about to exit delay timer activity
Rapid flashing = an alarm has occurred

CHIME MODE, {9} KEY

Chime mode alerts you to the opening of a perimeter door or window while the system is disarmed. When Chime mode is activated:

- three beeps sound whenever a perimeter door or window is opened

- interior zones (e.g., motion detector) do not produce a tone when they are triggered

- pressing the {*} key will display the faulted zones

- chime mode can be used only while the system is disarmed

- security code + {9}   The CHIME message appears when ON. Perimeter zones will cause a beep when faulted.

USING THE {OFF} OR {1} KEY

The {OFF} key is used to disarm the system, and silence the alarm.

- security Code + {1} key   The "READY" indicator light will be lit if all zones are secure, and the keypad will emit a single beep to confirm that the system is disarmed.

USING THE {BYPASS} OR {6} KEY

When bypassing zones:

- zone or zones can only be bypassed when the system is disarmed

- bypassed zones are unprotected zones

- when the system is disarmed, all zones are at default status (un-bypassed)

- security code + {6} key + zone number in single digit

- when finished, the LCD will momentarily display a "Bypass" message for each bypassed zone number. Arm the system as usual. When armed, the arming message is displayed as "ZONE BYPASSED."

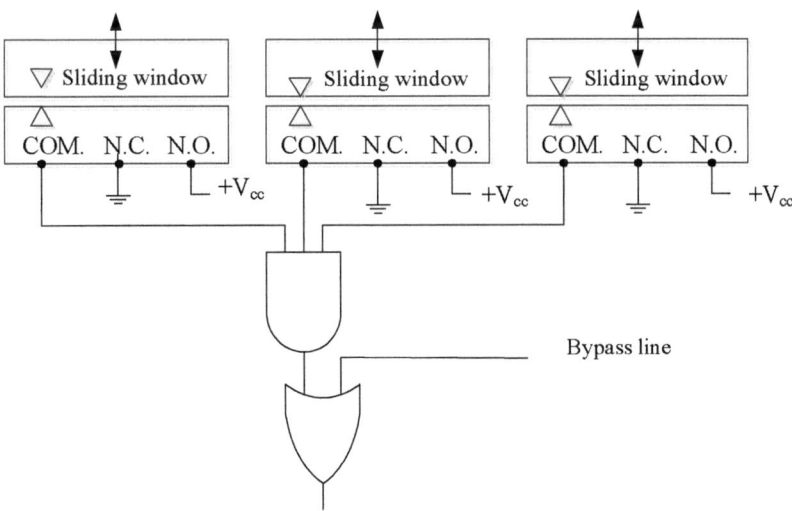

FIGURE 8.4:   Configuration to implement the bypass feature

Note: When the window is closed, N.O. is shorted to COM.

When the window is opened, N.C. is shorted to COM.

PROJECT #9:
Secondary PIN for the Home Security System

For most home security systems, there is a feature to assign a secondary PIN with the master or primary PIN. The secondary PIN is so the weekly cleaning person can have access to your house—but he or she will not have the capability to bypass any zone before arming the system.

For example, the homeowner activates the security system before he or she goes to work in the morning. Assuming all the zones are secure, including the motion detector, the cleaning team can enter the property with the secondary PIN, but can only activate the security system with the status assigned by the primary or master PIN earlier during the day.

In another scenario, if the homeowner has pets inside the house, he or she will activate the security system with the motion detector bypassed. When the cleaning team enters the property with the secondary PIN, the security system will get a snapshot of all the zones' status. When the cleaning team finishes the job, they can only activate the security system with the same status assigned by the primary or master PIN earlier during the day. No additional bypass is allowed.

This feature does not require any hardware; it is all controlled by software.

PART II
AUTONOMOUS ROBOTS

INTRODUCTION TO SERVO MOTORS

HOW TO CONTROL A SERVO MOTOR

HOW TO CHOOSE A CHASSIS

PROJECT #10: *Circular Motion*

PROJECT #11: *Figure Eight Configuration*

PROJECT #12: *Obstacle Avoidance*

PROJECT #13: *Navigate out of a Dead End*

PROJECT #14: *Running the Maze*

INTRODUCTION TO SERVO MOTORS

In general, the cost for a servo is between US$15 and $150. Regardless of how much you pay, there is no instruction manual to tell you how to control the purchased servo motor.[1] Most of the time, the answer is to search for videos on YouTube.

There are two types of servo motors: continuous rotation and 180° rotation.

Sometimes you might have to modify the servo for continuous rotation yourself. Some websites will charge a fee to convert the servo motor for continuous rotation.

The following instructions will guide you to convert a servo motor with metal gears for continuous rotation:

- Disassemble the servo until you can see all the gears.

- Use a dry marker (Sharpie) to mark a dot on the top of each gear.

- Carefully remove each gear without altering the shaft position.

- One of the driving gears has a pin as the stopper.

- Drill a tiny hole, slightly larger than 1/16", on a piece of wood.

- Put the driving gear on the piece of wood; place the pin on top of the tiny hole.

- Use a hammer and a center punch to knock off the pin.

- Carefully assemble all the driving gears back in the same fashion as before, with the marker on top, without altering the shaft position.

After converting the servo to continuous rotation, it is time to test the control mechanism for a servo motor.

In general, most of the servos have three wires: black, red, and yellow (HiTec); or brown, red, and orange. Different manufacturers use different colors for these three wires. The majority of them use red for the wire in the middle, usually for positive V_{cc}, and those two wires next to the red wire. The darker color is usually the ground or negative V_{cc}, and the brighter color is the input from the PWM port.

---

[1] http://www.seattlerobotics.org/guide/servos.html
http://www.jameco.com/Jameco/workshop/howitworks/how-servo-motors-work.html

HOW TO CONTROL A SERVO MOTOR

Equipment: Agilent E3630A triple output power supply

Agilent 33120A 15 MHz function arbitrary waveform generator

- Connect the +5 V to the red wire of the servo motor.
- Connect the −5 V to the black/brown wire of the servo motor.
- Connect the positive output of the function generator to the yellow/orange wire.
- Make sure the function generator and the power supply have a common ground.
- Set the waveform to be square wave.
- Set the peak-to-peak voltage on the function generator to 5 Vpp.
- Start checking the frequency of the square wave from 50 Hz and gradually up.
- Change the duty cycle of the square wave from 20% to 80%.
- Set the DC offset on the waveform to 2.5 V.

Using the waveform generator can eliminate all those steps to program your PWM from your development board.

The following tables II.1-II.3 illustrate how the duty cycle affects the pulse width related to the square wave frequency:

TABLE II.1

Freq. in Hz	Duty cycle	Pulse width in ms	Rotation	V_{DC}	DC offset
200	10%	0.50	0°	0.5 V	2.5 V
200	20%	1.00	0°	1.0 V	2.5 V
200	30%	1.50	90°	1.5 V	2.5 V
200	40%	2.00	180°	2.0 V	2.5 V
200	50%	2.50	180°	2.5 V	2.5 V
200	60%	3.00	180°	3.0 V	2.5 V
200	70%	3.50	180°	3.5 V	2.5 V
200	80%	4.00	180°	4.0 V	2.5 V

TABLE II.2

Freq. in Hz	Duty cycle	Pulse width in ms	Rotation	V_{DC}	DC offset
250	10%	0.40	0°	0.5 V	2.5 V
250	20%	0.80	0°	1.0 V	2.5 V
250	30%	1.20	0°	1.5 V	2.5 V
250	40%	1.60	90°	2.0 V	2.5 V
250	50%	2.00	180°	2.5 V	2.5 V
250	60%	2.40	180°	3.0 V	2.5 V
250	70%	2.80	180°	3.5 V	2.5 V
250	80%	3.20	180°	4.0 V	2.5 V

TABLE II.3

Freq. in Hz	Duty cycle	Pulse width in ms	Rotation	V_{DC}	DC offset
280	10%	0.36	0°	0.5 V	2.5 V
280	20%	0.71	0°	1.0 V	2.5 V
280	30%	1.07	0°	1.5 V	2.5 V
280	40%	1.43	90°	2.0 V	2.5 V
280	50%	1.79	90°	2.5 V	2.5 V
280	60%	2.14	180°	3.0 V	2.5 V
280	70%	2.50	180°	3.5 V	2.5 V
280	80%	2.86	180°	4.0 V	2.5 V

From the above three tables, you probably can tell the frequency range (250 Hz ~ 300 Hz) that you want to generate with your PWM. For example, when frequency is at 280 Hz, with duty cycle at 40%, then the rotation is at neutral (90°). The transition from a neutral position to 0° will require the duty cycle to drop down to 30% or lower. Whether we want to classify this as the counter-clockwise direction will depend on the reference point. No matter what, the transition from neutral position to 180° will require the duty cycle to increase. The rotation is just the opposite of the rotation from neutral position to 0°. Only the duty cycle can affect the DC voltage of the waveform, and the DC voltage is controlling the RPM of the servo motor.

Now that we have introduced the IR sensor and servo motor control, it is time to discuss the robot chassis kit. In any autonomous robotic competition, they usually announce the size of the robot. For example, the robot before stretching out should be able to fit into a 1 foot × 1 foot × 1 foot cube. It all depends on all those criteria that the competition requires each robot to perform. Here, we will focus more on education and cover most of the basic features for most robotic competition.

The recommended kits are:
http://www.budgetrobotics.com

- Arduino Robot Bonanza Teachbot Chassis Kit

- ArdBot Chassis Kit

- ArdBot II

All three of these chassis kits are similar; the only difference will be the deck area.

We have been using the Teachbot Chassis Kit for my robotics course, and we use Dragon12-JR by EVBplus (Wytec Company). Dragon12-JR is the development board for our Microprocessor Systems course. Students are taught to program the microprocessor 9S12 in assembly with AsmIDE v3.40.

HOW TO CHOOSE A CHASSIS

In general, all competitions have a size constraint for robots. The type of material for the chassis will depend on the constraint with weight. If there is no restriction on the material, then the best bet is to use lighter material. When students build their robots for competition or senior projects, most of them use an Arduino development board, but most faculty will not use Arduino as a teaching tool. Programming a robot is mainly input/output programming; whether you are using C or Object Orientated C, you still need to know how to set those input/output PINs. The major difference in C is the *FOR* loop and *IF* statement. In assembly you have to implement the *FOR* loop and use the eight flags in the Conditional Code Register to implement the *IF* statement. Most of the programming for the robot is to change the input or output of certain PORT, and there is no number crunching. Instead, it might keep on counting like the wheel watcher or the output-compare function for time delay.

Once you decide on the chassis, it is time to look into the servo and wheel or tire set. What type of servo motor you need will depend on the surface that the robot will perform on. The cheap servo motor usually has plastic gears and is designed to run on a smooth surface. If you need the robot to run over some rough or uneven surfaces, then it may be a good idea to get a high-torque metal gear servo motor with continuous rotation. Certain types of servo motor can only fit certain types of wheel. I strongly recommend looking into the Nubotics WW-11 Wheel Watcher.[2] The angular rotation of a servo motor needs some type of feed-

---

2   http://www.nubotics.com/products/ww11/index.html

back to allow the microprocessor to keep track of the rotation. In general, the microprocessor development board can power up with a 9 V battery, but the two servo motors will definitely need another battery pack that can supply at least 1000 mAh or more. The RPM of a servo motor is dependent on the DC voltage in the battery pack. Since the robot is running on the DC voltage from the batteries, after testing the robot for 15 minutes, the voltage in the battery pack can easily drop down by 0.5 V or more. The RPM of the servo will also drop, thus the timer cannot provide an accurate measurement of the distance. Instead we need the feedback data from the wheel watcher to provide the feedback to do the job. Getting an extra battery pack and a charger are essential. While you are testing your robot, you can use the charger to charge up the extra battery pack.

With the addition of a wheel watch like WW-11, you then have to choose the wheel that fits your servo motor, and also has a flat surface on the inside of the wheel/tire for the self-adhesive codewheel.[3] I modified the black disc that came with the HiTec HS-645MG. I used a mini-table drill to drill four holes on the black disc, and countersunk the drilled hole on the inside of the black disc. Later I bolted the black disc to the wheel or tire with four countersunk screws, and locked it with four nuts on the external surface of the tire. Make sure the countersunk screws are flush with the surface of the black disc. Then the self-adhesive codewheel can stick to the inside surface of the black disc and you can insert the screw to attach the wheel to the servo motor.

Last but not least, I replaced the deck riser with four piece of 1"×2" wood, and used eight wooden screws to screw through the holes on the lower and upper deck as for the riser. Although the riser from the kit is 2.5" high, you can cut the wood to 2" instead. According to the Sharp IR sensor data sheet, they recommend mounting the IR sensor vertically. Those four pieces of wood will be perfect to mount each IR sensor vertically with two tiny screws. More wooden posts will be needed if more IR sensors are necessary.

---

[3] http://www.nubotics.com/products/ww11/index.html

PROJECT #10:
Circular Motion

Program your robot to run a circle with a radius of 24 inches.

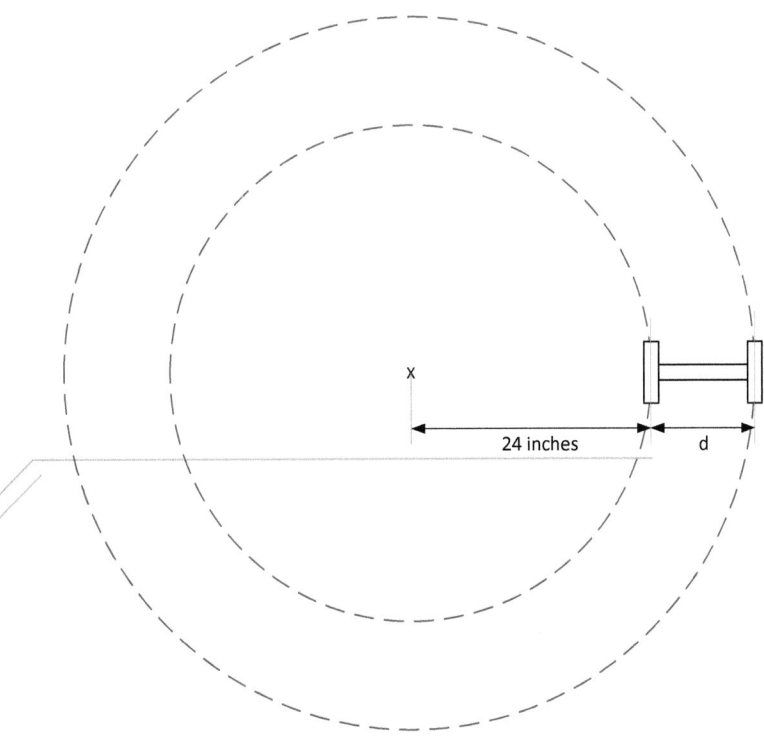

FIGURE 10.1: Circular motion for an autonomous robot

Use a tape measure to measure the circumference of the wheel with a rubber tire.

The circumference of the inside circle   = $2\pi \times 24$ inches

The circumference of the outside circle   = $2\pi \times (24 + d)$ inches

Use the PWM to control the RPM of each wheel. The outside wheel should rotate at a higher RPM than the inside wheel. Take two pieces of masking tape, and stick them to the floor 48 inches or four tiles apart. Start your robot with the inside wheel aligned with one piece of the masking tape. After running one full circle, the inside tire of your robot should run over the other piece of tape and then stop on the first piece as it was at the starting point.

For an autonomous robot, there is no axil, and the RPM of each servo is controlled individually. The two servos are mounted with their backs facing each other. For any forward movement, the servo on the left is turning **counterclockwise** (with the bottom deck facing the ground as the reference), the servo on the right will be making a **clockwise rotation** (with the bottom deck facing the ground as the reference). If you take the imaginary center line of the circular deck, and fold the two halves together, what you see will be two servos on top of each other, as shown in Figure 10.2.

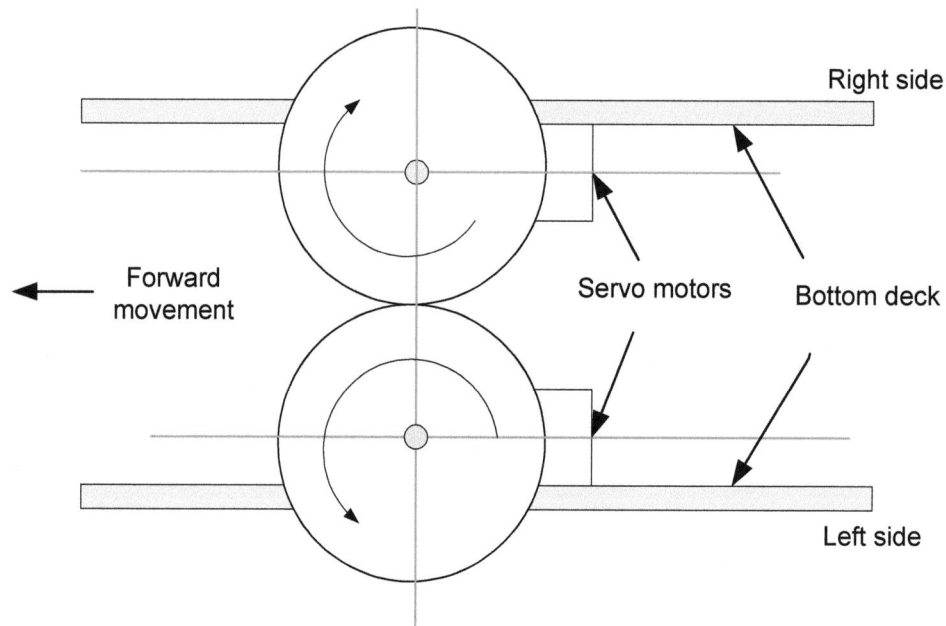

FIGURE 10.2: Illustration of the rotation of the two servo motors

First you have to select a frequency with a duty cycle that gets the servo motor running in a counterclockwise rotation; for example, 250 Hz square wave with 20~30% duty cycle for the left servo. You can go ahead and use PWM to implement the code for the left servo, and then you can use trial and error to implement the code for the right servo. Keep in mind that the right servo motor should run at a higher RPM in its clockwise rotation than the left servo motor in its counterclockwise rotation. Visual inspection of the servo rotation with the arbitrary waveform generator for various duty cycles will shorten your trial-and-error approach.

Use the wheel watcher to control one circular run and stop after one circle.

PROJECT #11:
Figure Eight Configuration

Program your robot to run a circle with a radius of 24 inches.

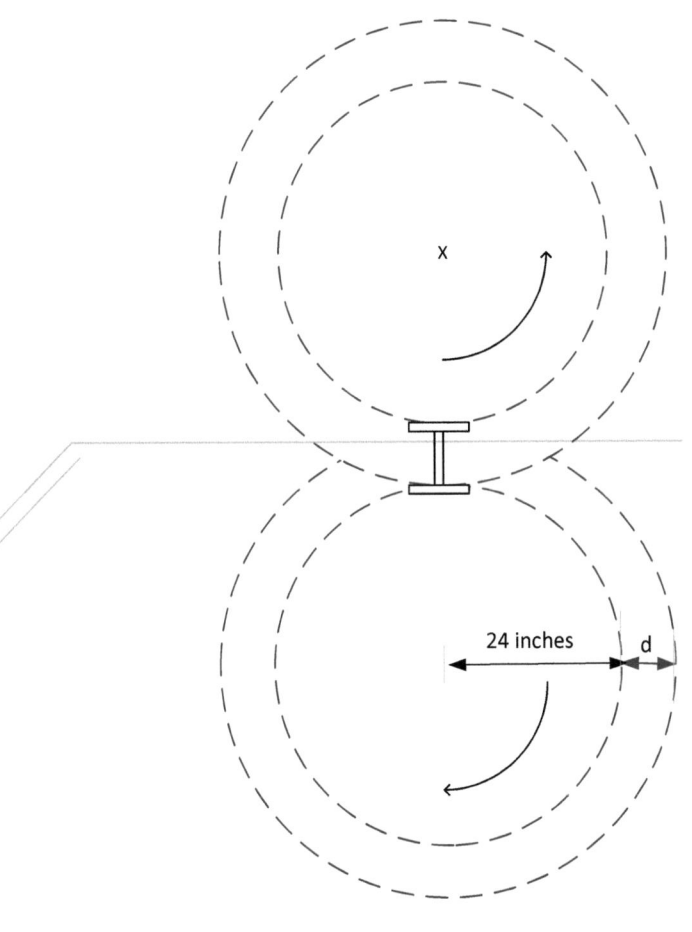

FIGURE 11.1: The formation for running a figure eight

Use a tape measure to measure the circumference of the wheel with a rubber tire.

The circumference of the inside circle = $2\pi \times 24$ inches

The circumference of the outside circle = $2\pi \times (24 + d)$ inches

Use the PWM to control the RPM of each wheel. The outside wheel should rotate at a higher RPM than the inside wheel. Take two pieces of masking tape and stick them to the floor 48 inches or four tiles apart. Start your robot with the inside wheel aligned with one piece of the masking tape. After running one full circle, the inside tire of your robot should run over the other piece of tape and then stop on the first piece as it was at the starting point. Use the wheel watcher encoder to control the circle run.

Programming the robot to run a digit eight configuration requires a high degree of accuracy in running a complete circle. Let the robot pause for a few seconds before running another circle in the opposition direction. The robot can run the top circle in a counterclockwise direction, and when it returns to the starting point it should pause before continuing the lower circle in clockwise direction, or vice versa. The vertical axis of the digit eight will drift, too.

PROJECT #12:
Obstacle Avoidance

For all three chassis kits from budgetrobotics.com, the lower deck has various holes, and those holes on the outskirt are inclined at 45° from each other. If you pick the Teachbot Chassis Kit, you can see all eight holes on the outskirt of the upper deck. You can pick the three holes up front on the outskirt of the upper deck to mount the wooden post, and the lower deck also has three holes align with the upper deck. The three wooden posts will support the IR sensors for obstacle avoidance, and making decisions for left or right turns. Sharp GP2D120 or GP2D12 should be able to handle the short range of about 5 or 6 inches. How you mount the two sensors on the right and left will depend on your algorithm to make the turn. The maze will be built with 1 × 4 (actual dimension is 3/4″ × 3.5″).

Program your robot such that it will stop about 6 inches ahead of the obstacle. Most of pathway within the maze is about 14 inches, and all the turns within the maze are 90°. The decision to make a left turn is made when both the mid and right IR sensors see an obstacle ahead. The decision to make a right turn is made when the mid and left IR sensors see an obstacle ahead.

How many ways are there to control the two servos in order to make a left turn?

- The left servo is idle, and the right servo motor is moving forward. If the running surface is tile, the left tire may skip.

- The left servo rotates at a lower RPM, and the right servo rotates at a higher RPM.

- The left servo moves backward at a certain RPM, and the right servo moves forward in the same RPM as the left servo. That means the robot will pivot with respect to the center of the circular deck. This method will be the best in a tight space.

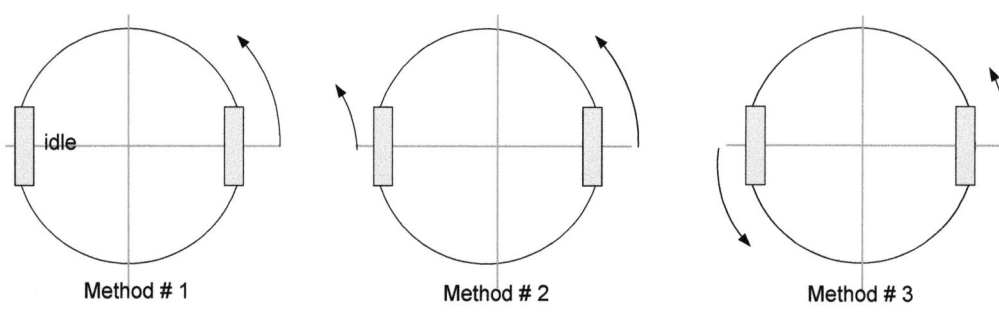

FIGURE 12.1: Three different methods for making a left turn

Note: The best way to monitor the turn is by reading the counts through the codewheel of the Wheel Watcher WW-11.

PROJECT #13:
Navigate out of a Dead End

With all three IR distance sensors, program your robot to find its way out from a dead end. Two dead end configurations are shown in Figures 13.1a and 13.1b.

The internal dimension of the hallway will be **at least** 14 inches wide.

The tentative dimension of the maze is about 6 feet by 6 feet.

The depth of a dead end can be more than 14 inches.

All the turns are 90°.

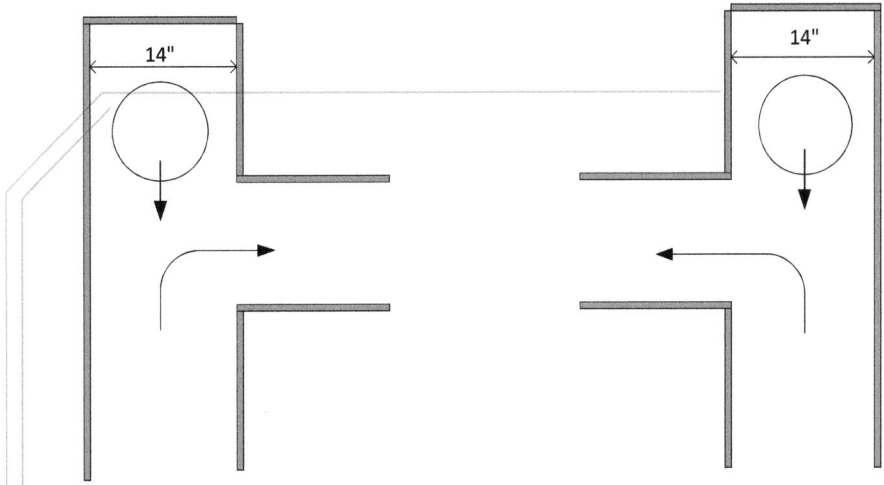

FIGURE 13.1A-B

13.1A: DEAD END configuration for a right exit   **13.1B:** Dead end configuration for a left exit

There is one dead end in the maze. Students have to program their robots to find a way out from a dead end situation. Since GP2D120 can sense any obstacle up to 12 inches, GP2D12 may perform better in a dead end scenario, because the range for GP2D12 can go up to 30 inches.

With Sharp GP2D120, your robot will sense the dead end when it gets into it. The only way out is to back up until the left or right sensor senses an opening and turns to the opening.

With Sharp GP2D12, your robot can sense the wall from the mid IR sensor while the robot is at the intersection. The backup procedure may not be necessary.

PROJECT #14:
Running the Maze

a. Your autonomous robot should be able to run through the maze from entrance to exit.

b. Your autonomous robot should be able to run through the maze from exit to entrance.

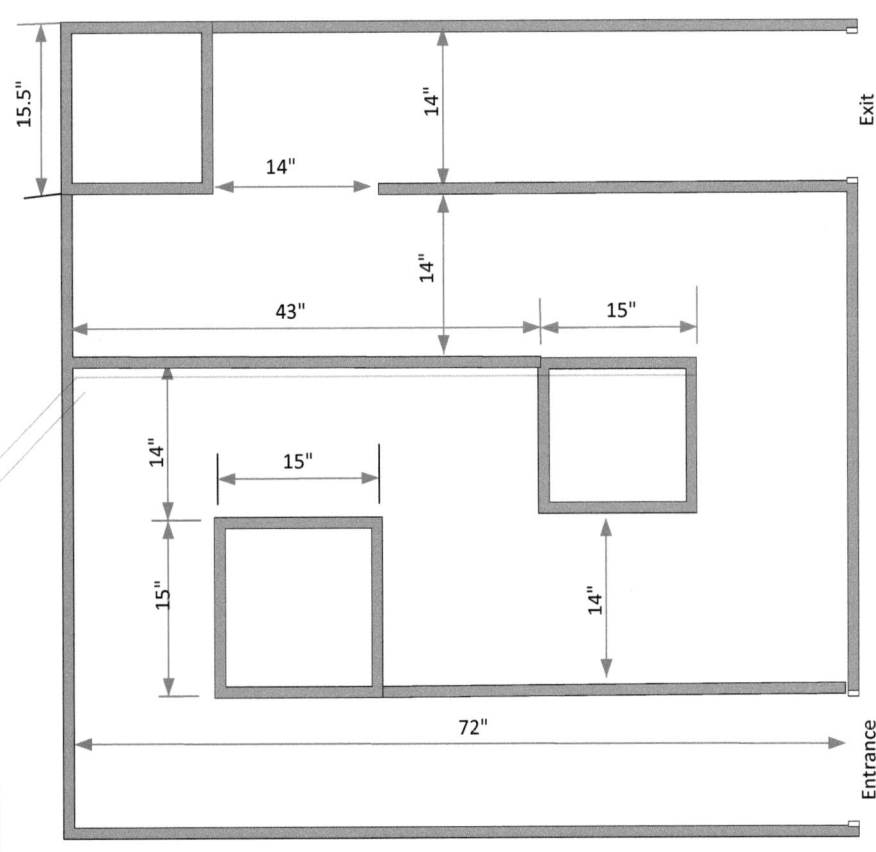

FIGURE 14.1: Maze configuration

Students should program their robots to perform two runs through the maze. During these two runs, the robot should be controlled by **one** program. Those two runs are specified as part a) and part b). In order to strengthen the maze, I added the return path to link the exit and the entrance. The additional overpasses are used to keep the maze in tack, because the high torque HiTec HS-645MG can push the board by itself, and change the maze configuration with its desire. Students will take at least one or two weeks to fine-tune their robots before the competition. The instructor can implement the motion-triggered sensor Sharp GP2D15 to activate the timer at the entrance, and another GP2D15 to stop the timer at the exit. If the robot is running the path for part b), just swap the two pins for the two GP2D15 within the I/O Port. The electronic time lapse for each run will be recorded. The fastest time to execute path a) and b) will be the winner of the race.

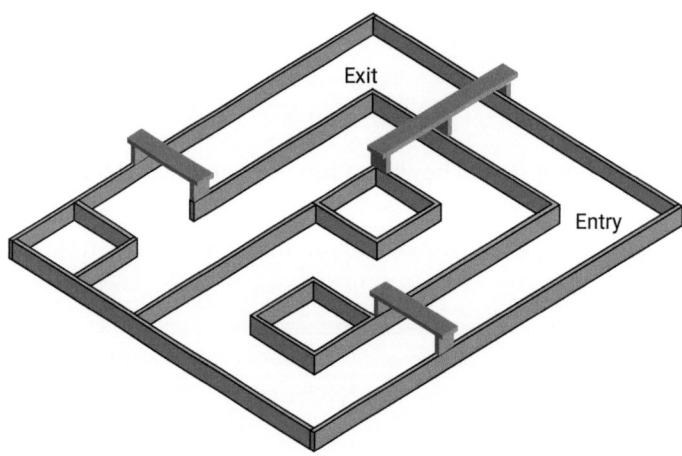

FIGURE 14.2: Isometric view of the maze

Note: The enclosure is to keep the robot running through the maze continuously (for exhibition purposes). The bridges are to reinforce the pathways being pushed by the robot.

PART III
MICROPROCESSOR LABS

LAB #1   *Addressing Mode*

LAB #2   *Compute the Array Sum*

LAB #3   *Condition Flags and Rotate Instruction*

LAB #4   *Bit Testing*

LAB #5   *Bit Manipulation*

LAB #6   *LED Traffic Light*

LAB #7   *PIN Verification with 4×4 Matrix Keypad*

LAB #8   *PIN Verification with Single Pole/Common Bus Keypad*

LAB #9   *Siren Generation*

LAB #10 *Dim an LED*

LAB #11 *Home Security System*

LAB #1:
Addressing Mode

a) Type in the following assembly program:

```
ORG     $2000 ; starting address is @ $2000
LDD     #$BBCC ; load constant $BBCC to Register D
STAA    $1501 ; store accumulator A to location $1501
STAB    $1502 ; store accumulator B to location $1502
LDAA    $1501 ; [A] <= [$1501]
LDAB    $1502 ; [B] <= [$1502]
STOP
END
```

With the AsmIDE, once you assemble the program, you will get a *filename.lst* file. Open the list file, and you will see the following:

as12, an absolute assembler for Motorola MCU's, version 1.2h

```
2000            ORG $2000 ;starting address is @ $2000
2000  cc  bb    ccLDD      #$BBCC ;load constant $BBCC to Register D
2003  7a  15    01STAA     $1501 ;store accumulator A to location $1501
2006  7b  15    02STAB     $1502 ;store accumulator B to location $1502
2009  b6  15    01LDAA     $1501 ;[A] <= [$1501]
200c  f6  15    02LDAB     $1502 ;[B] <= [$1502]
200f  18  3e    STOP
END
```

Executed: Sun Sep 13 22:09:52 2015
Total cycles: 28, Total bytes: 17
Total errors: 0, Total warnings: 0

The first column is for the Program Counter (PC).

Reset the Dragon12-JR and load the program to the development board.

Type br 2006. That is to set the breakpoint at address $2006.

Type g 2000, and you will see the following:
User Bkpt Encountered

```
PP  PCSPXY  D = A:B  CCR = SXHI NZVC
3820063C000000 0000BB: CC1001 1000
XX: 20067B1502 STAB $1502
```

At the cursor, type md 1500

Note: md stands for memory display.

The display will be as follows:
1500 B9 BB FE 47 – 0E EB 4B 9E – 64 59 00 7D – E7 8F 34 22

The second number represents the address location $1501, the content in $1501 is $BB.

The instruction   STAA $1501   ; [$1501] <= [A]

Similarly, you can set another breakpoint at location $2009 by typing br 2009, and checking the content in memory location $1502.

b) The following assembly program illustrates the different types of indexed addressing:

```
        ORG     $2000           ;starting address is @ $2000
;loading data to the memory
        LDD     #$0011                  ;load constant $0011 to Acc. A:B
        STAA    $1500                   ;[$1500] = $00
        STAB    $1501                   ;[$1501] = $11
        LDD     #$2233                  ;load constant $2233 to Acc. A:B
        STAA    $1502                   ;[$1502] = $22
        STAB    $1503                   ;[$1503] = $33
        LDD     #$4455                  ;load constant $4455 to Acc. A:B
        STAA    $1504                   ;[$1504] = $44
        STAB    $1505                   ;[$1505] = $55
        LDD     #$6677                  ;load constant $6677 to Acc. A:B
        STAA    $1506                   ;[$1506] = $66
        STAB    $1507                   ;[$1507] = $77
        LDD     #$8899                  ;load constant $8899 to Acc. A:B
        STAA    $1508                   ;[$1508] = $88
        STAB    $1509                   ;[$1509] = $99
        LDD     #$AABB                  ;load constant $AABB to Acc. A:B
        STAA    $150A                   ;[$150A] = $AA
        STAB    $150B                   ;[$150B] = $BB
```

```
        LDD         #$CCDD                      ;load constant $CCDD to Acc. A:B
        STAA        $150C                       ;[$150C] = $CC
        STAB        $150D                       ;[$150D] = $DD
        LDD         #$EEFF                      ;load constant $EEFF to Acc. A:B
        STAA        $150E                       ;[$150E] = $EE
        STAB        $150F                       ;[$150F] = $FF
;5-bit Constant Offset Indexed Addressing ( −16 ~ 15)
        LDX         #$1509
        LDAA        5,X                         ;$1509 + 5 = $150E
        LDAB        −8,X                        ;$1509 − 8 = $1501
;9-bit Constant Offset Indexed Addressing (offset is −256 ~ +255)
        LDY         #$1590
        LDAA        −140,Y                      ;$1590 − $8C = $1504 (note: 140 ≡ $8C)
        LDY         #$1490
        LDAB        $76,Y                       ;$1490 + $76 = 1506
;16-bit Constant Offset Indexed Addressing
        LDX         #$1000
        LDAA        $50F,X
        LDY         #$2000
        LDAB        −$AF6,Y                     ;$2000 − $AF6 = $150A
;16-bit Constant Indirect Indexed Addressing
        LDD         #$1505                      ;[A] = $15 & [B] = $05
        STAA        $1515                       ;$1515 <= $15
        STAB        $1516                       ;$1516 <= $05
        LDX         #$152D
        LDAA        [−$18,X]                    ;$152D − $18 = $1515 & [$1515] = $15
;16-bit Constant Indirect Indexed Addressing
        LDD         #$1508                      ;[A] = $15 & [B] = $08
        STAA        $1545                       ;[$1545] <= $15
        STAB        $1546                       ;[$1546] = $08
        LDX         #$152D
        LDAB        [$19,X]                     ;$152D+$19=$1546 & [$1546] = $08
        STOP
        END
```

At the end of each Offset Index Addressing, set your breakpoint. Verify the content in either Accumulator A or B.

Keep in mind the address content for md 1500 is declared in the beginning as follows:
1500   00 11 22 33 − 44 55 66 77 − 88 99 AA BB − CC DD EE FF

Modify the offset for each type of indexed addressing, and verify and record two different offsets for each type.

c)   Modify the loading data to memory for the above assembly program. Use Auto Pre/Post-Increment Indexed Addressing; e.g., LDAA   1,Y+

Verify the contents in each of the address locations.

You can use the provided registers map to illustrate the content in certain memory locations.

Registers Map

Register	Value
A	
D	
X	
Y	
SP	
PC	

S	X	H	I	N	Z	V	C

B

Address	
$1500	$00
$1501	$11
$1502	$22
$1503	$33
$1504	$44
$1505	$55
$1506	$66
$1507	$77
$1508	$88
$1509	$99
$150A	$AA
$150B	$BB
$150C	$CC
$150D	$DD
$150E	$EE
$150F	$FF

Address	
$1510	
$1511	
$1512	
$1513	
$1514	
$1515	
$1516	
$1517	
$1518	
$1519	
$151A	
$151B	
$151C	
$151D	
$151E	
$151F	

Address	
$1520	
$1521	
$1522	
$1523	
$1524	
$1525	
$1526	
$1527	
$1528	
$1529	
$152A	
$152B	
$152C	
$152D	
$152E	
$152F	

Register	Value
A	
D	
X	
Y	
SP	
PC	

S	X	H	I	N	Z	V	C

B

Address	
$1500	$00
$1501	$11
$1502	$22
$1503	$33
$1504	$44
$1505	$55
$1506	$66
$1507	$77
$1508	$88
$1509	$99
$150A	$AA
$150B	$BB
$150C	$CC
$150D	$DD
$150E	$EE
$150F	$FF

Address	
$1510	
$1511	
$1512	
$1513	
$1514	
$1515	
$1516	
$1517	
$1518	
$1519	
$151A	
$151B	
$151C	
$151D	
$151E	
$151F	

Address	
$1520	
$1521	
$1522	
$1523	
$1524	
$1525	
$1526	
$1527	
$1528	
$1529	
$152A	
$152B	
$152C	
$152D	
$152E	
$152F	

LAB #2:
Compute the Array Sum

For Example 2.14 in your text, change the number of elements in the array to 4, and the array declared as 10, 13, 17, and 20.

Please do not overwrite the old version; keep all versions for review.

- a. Determine the intermediate sum within the loop for all four iterations and record the data with the provided registers map.

- b. Initialize the loop index i to be 4, and decrement i as it goes through the loop. Modify the criterion for the program to stop, for $i = 4$ down to 1.

- c. Repeat a) and b) such that the loop will compute the sum by adding the last element first, as $0 + 20$ in the first iteration, and then perform sum = $20 + 17$, et cetera. Add the elements backwards.

- d. Modify your assembly program such that the array is located at $1500 ~ $1504 before org 2000.

- e. Repeat parts b) and c) with the array declaring in the beginning of the assembly program.

```
N      equ    4
       org    $1500
sum    rmb    2
i      rmb    1
       org    $2000
       ldaa   #0              ; A = 0
       staa   i               ; i = 0
       staa   sum             ; sum(MSB) ← 0
       staa   sum+1           ; sum(LSB) ← 0
loop   ldab   i               ; B = i
```

```
            cmpb      #N                      ; is B =20 or N?
            beq       done                    ; if equal, branch to done
            ldx       #array                  ; X = address of array
            abx                               ; X = B + X
            ldab      0,X                     ; B = [x]
            ldy       sum                     ; Y = sum
            aby                               ; Y = B + Y
            sty       sum                     ; sum = Y
            inc       i                       ; increment the loop count by 1
            bra       loop                    ; branch back to loop
      done  swi                               ; software interrupt
      array dc.b      10,13,17,20             ; declare the array elements
            end
```

Registers Map

A			B
D			

X	

Y		

SP	

PC	

S	X	H	I	N	Z	V	C

Address	
$1500	$00
$1501	$11
$1502	$22
$1503	$33
$1504	$44
$1505	$55
$1506	$66
$1507	$77
$1508	$88
$1509	$99
$150A	$AA
$150B	$BB
$150C	$CC
$150D	$DD
$150E	$EE
$150F	$FF

Address	
$1510	
$1511	
$1512	
$1513	
$1514	
$1515	
$1516	
$1517	
$1518	
$1519	
$151A	
$151B	
$151C	
$151D	
$151E	
$151F	

Address	
$1520	
$1521	
$1522	
$1523	
$1524	
$1525	
$1526	
$1527	
$1528	
$1529	
$152A	
$152B	
$152C	
$152D	
$152E	
$152F	

				B		Address		Address		Address
A					$1500	$00	$1510		$1520	
D					$1501	$11	$1511		$1521	
					$1502	$22	$1512		$1522	
X					$1503	$33	$1513		$1523	
					$1504	$44	$1514		$1524	
Y					$1505	$55	$1515		$1525	
					$1506	$66	$1516		$1526	
SP					$1507	$77	$1517		$1527	
					$1508	$88	$1518		$1528	
PC					$1509	$99	$1519		$1529	
					$150A	$AA	$151A		$152A	

S	X	H	I	N	Z	V	C

Address	
$150B	$BB
$150C	$CC
$150D	$DD
$150E	$EE
$150F	$FF

Address	
$151B	
$151C	
$151D	
$151E	
$151F	

Address	
$152B	
$152C	
$152D	
$152E	
$152F	

LAB #3:
Condition Flags and Rotate Instruction

Use Appendix A in your text to determine the operation for the following **opcodes**.[1]

Use the registers map to identify the register content.

a. Find the values of condition flags N, Z, V, and C in the CCR register after the execution of each of the following instructions, given that **[A] = $50** and the condition flags are **N = 0, Z = 1, V = 0, and C = 1**. Illustrate each instruction in your own words.

CCR	S	X	H	I	N	Z	V	C
$05	x	x	x	x	0	1	0	1

where x stands for don't care.

```
ORG     $1500
ORCC    #$05            ; logical OR CCR with constant $05
                        ; set Z & C bits
ANDCC   #$F5            ; logical AND CCR with constant $F5
                        ; clear N & V bits
LDAA    #$50            ; A <= %01010000 or $50
SUBA    #40             ; A <= A − $28, 40 ≡ $28
TSTA
ADDA    #$50
LSRA
ROLA
LSLA
```

---

[1] Han-Way Huang, *HCS12/9S12 An Introduction to Software and Hardware Interfacing*, 2nd ed. (Delmar Cengage Learning, 2010).

i) ORCC #$05

	S	X	H	I	N	Z	V	C
CCR	x	x	x	x	x	x	x	x
$05	0	0	0	0	0	1	0	1
ORCC	x	x	x	x	x	1	x	1

The OR function sets Z and C bits, and x stands for don't care.

ii) ANDCC #$F5

	S	X	H	I	N	Z	V	C
CCR	x	x	x	x	x	1	x	1
$F5	1	1	1	1	0	1	0	1
ANDCC	x	x	x	x	0	1	0	1

The AND function clears N and V bits, and x stands for don't care.

b. Find the values of condition flags N, Z, V, and C in the CCR register after the execution of each of the following instructions **independently**, given that [A] = $00 and the initial condition codes are N = 0, Z = 1, V = 0, and C = 0. Illustrate each instruction in your own words.

CCR	S	X	H	I	N	Z	V	C
$05	x	x	x	x	0	1	0	0

where x stands for don't care.

```
ORG         $1500
ORCC        #$04            ;logical OR CCR with constant $04
                            ;sets Z bit

ANDCC #$F4                  ;logical AND CCR with constant $F4
                            ;clears N, V & C bits

LDAA        #$50            ;A <= %01010000 or $50
SUBA        #40             ;A <= A − $28 , 40 ≡ $28
TSTA
ADDA        #$40
SUBA        #$78
LSLA
ROLA
ADDA        #$CF
```

i) ORCC #$04

	S	X	H	I	N	Z	V	C
CCR	x	x	x	x	x	x	x	x
$04	0	0	0	0	0	1	0	0
ORCC	x	x	x	x	x	1	x	x

The OR function sets Z bit.

ii) ANDCC #$F4

	S	X	H	I	N	Z	V	C
CCR	x	x	x	x	x	1	x	x
$F4	1	1	1	1	0	1	0	0
ANDCC	x	x	x	x	0	1	0	0

The AND function clears N, V, and Z bits.

c. Assemble and execute the following program for multi-precision addition.

$A2958287 + $$78458998

```
ORG   $2000
LDD   #$8287
ADDD  #$8998
STD   $1502
LDAA  #$95
ADCA  #$45
STAA  $1501
LDAA  #$A2
ADCA  #$78
STAA  $1500
END
```

Use the registers map to identify the register content.

Single step each instruction, and pay attention to the value of the C flag.

d. Assemble and execute the following program for multi-precision subtraction.

$87654321 − $23456789

```
ORG   $2000
LDD   #$4321
SUBD  #$6789
```

```
STD     $1502
LDAA    #$65
SBCA    #$45
STAA    $1501
LDAA    #$87
SBCA    #$23
STAA    $1500
END
```

Use the registers map to identify the register content.

Single step each instruction, and pay attention to the value of the C flag.

Registers Map

A		
D		
X		
Y		
SP		
PC		

S	X	H	I	N	Z	V	C

Address		Address		Address	
$1500	$00	$1510		$1520	
$1501	$11	$1511		$1521	
$1502	$22	$1512		$1522	
$1503	$33	$1513		$1523	
$1504	$44	$1514		$1524	
$1505	$55	$1515		$1525	
$1506	$66	$1516		$1526	
$1507	$77	$1517		$1527	
$1508	$88	$1518		$1528	
$1509	$99	$1519		$1529	
$150A	$AA	$151A		$152A	
$150B	$BB	$151B		$152B	
$150C	$CC	$151C		$152C	
$150D	$DD	$151D		$152D	
$150E	$EE	$151E		$152E	
$150F	$FF	$151F		$152F	

A [___][___] B

D [_____]

X [_____]

Y [___][___]

SP [_____]

PC [_____]

S	X	H	I	N	Z	V	C

Address	
$1500	$00
$1501	$11
$1502	$22
$1503	$33
$1504	$44
$1505	$55
$1506	$66
$1507	$77
$1508	$88
$1509	$99
$150A	$AA
$150B	$BB
$150C	$CC
$150D	$DD
$150E	$EE
$150F	$FF

Address	
$1510	
$1511	
$1512	
$1513	
$1514	
$1515	
$1516	
$1517	
$1518	
$1519	
$151A	
$151B	
$151C	
$151D	
$151E	
$151F	

Address	
$1520	
$1521	
$1522	
$1523	
$1524	
$1525	
$1526	
$1527	
$1528	
$1529	
$152A	
$152B	
$152C	
$152D	
$152E	
$152F	

LAB #4:
Bit Testing

a. With reference to Example 2.23 in your text, write a program to count the number of 1s contained in memory locations $1500~$1501 and save the result in memory location $1505.[1]

b. With reference to Example 2.17 in your text, write a program to sort out the number that is **divisible by four** and the number that is **not divisible by four**. Assuming that the starting address of your program is at $2000, the first number that is divisible by four will be stored at memory location at $1500, the second number will be at $1501, and so on. The first number that is not divisible by four will be stored in memory location at $1520, the second number will be at $1521, et cetera. You can keep the same twenty numbers of the array as follows:

    array   db   1, 3, 5, 6, 19, 41, 53, 28, 13, 42, 76, 14, 20,
    54, 64, 74, 29, 33, 41, 45

c. By checking the remainder with *idiv*, write a program to sort out the **number that is divisible by three**, and **not divisible by three**. Assuming that the starting address of your program is at $2000, the first number that is divisible by three will be stored at memory location at $1500, and so on. The first number that is not divisible by three will be stored in memory location at $1510, et cetera. The array is defined as the following:

    array   db   1, 3, 5, 6, 19, 41, 53, 28, 13, 42, 76, 14, 20,
    54, 64, 74, 29, 33, 41, 45

Use integer division to determine the remainder. If the remainder is equal to 0, that means the number is divisible by three. As a matter of fact, this approach can also be applied to part b.

---

[1] Han-Way Huang, *HCS12/9S12 An Introduction to Software and Hardware Interfacing*, 2nd ed. (*Delmar Cengage Learning, 2010*).

LAB #5:
Bit Manipulation

Assign #$5A or #%01011010 to memory location $1500, and load the memory location to Accumulator A or B to perform the bit manipulation. Then store Accumulator A or B back to the same memory location.

a. Write an instruction sequence to **toggle** bits 3, 2, 1, and 0 of memory location at $1500 and **clear** the other four bits (7, 6, 5, and 4) from the same location. Provide a comment statement for each instruction.

b. Write an instruction sequence to **toggle** bits 7, 6, 5, and 4 of memory location at $1500 and let the other four bits (3, 2, 1, and 0) remain **unchanged** at the same location. Provide a comment statement for each instruction.

c. Write an instruction sequence to **toggle** bits 7, 6, 5, and 4 of memory location at $1500 and **set** the other four bits (3, 2, 1, and 0) at the same location. Provide a comment statement for each instruction.

d. Write an instruction sequence to **toggle** bits 3, 2, 1, and 0 of memory location at $1500 and **set** the other four bits (7, 6, 5, and 4) at the same location. Provide a comment statement for each instruction.

e. Write an instruction sequence to **set** bits 7, 5, 3, and 1 of a memory location and **clear** the other four bits at the same location. Provide a comment statement for each instruction.

f. Write an instruction sequence to **swap** the **upper four bits** with the **lower four bits** of accumulator **B**. (Swap bit 7 with bit 3, bit 6 with bit 2, and so on.)

g. Write an instruction sequence to **swap** the **upper four bits** with the **lower four bits** of accumulator **A**. (Swap bit 7 with bit 3, bit 6 with bit 2, and so on.)

bit	7	6	5	4	3	2	1	0
$5A	0	1	0	1	1	0	1	0

LAB #6:
LED Traffic Light

Use the HCS12 Port B to drive **three** LEDs (green, yellow, and red). Light each of them for half a second in sequence and repeat, assuming that the Dragon12-JR is running at 24 MHz E clock.

Use the timer to generate the 1 second delay time first. Repeating the 1 second delay for three times with a loop will generate the 3 second delay.

Modify the above assembly program to simulate the sequence of a traffic light.

a. Light up the green and red LEDs for 3 seconds, and the yellow LED for 1 second.

b. With the prescale factor set to 32, and use OC3 for output compare.

c. Light up the green and red LEDs for 3 seconds, and the yellow LED for 1 second, then light up the green and red LEDs for 5 seconds, and the yellow LED for 1 second. Repeat the sequence again and again (nonstop).

```
;the following subroutine creates a 10 ms delay with the E clock at 24 MHz
delay10ms   movb    #$90,TSCR1          ;enable TCNT and fast flags clear
            movb    #$06,TSCR2          ;configure prescale factor to 64
            movb    #$01,TIOS           ;enable OC0
            ldd     TCNT
            addd    #3750               ;start an output compare operation
            std     TC0                 ;with 10 ms time delay
wait        brclr   TFLG1,$01,wait      ;if equal, C0F in TFLG1 is set to 1
            rts
```

Note: You can also remove the label wait with the following:
```
            brclr   TFLG1,$01,*         ;the asterisk means itself
```

Registers related to the output-compare function:

	7	6	5	4	3	2	1	0	
$0040	IOS7	IOS6	IOS5	IOS4	IOS3	IOS2	IOS1	IOS0	TIOS
$004D	TOI	0	0	0	TCRE	PR2	PR1	PR0	TSCR2
$0044	Bit 15	14	13	12	11	10	9	Bit 8	TCNT(H)
$0043	Bit 7	6	5	4	3	2	1	Bit 0	TCNT(L)
$0046	TEN	TSWAI	TSFRZ	TFFCA	0	0	0	0	TSCR1
$004E	C7F	C6F	C5F	C4F	C3F	C2F	C1F	C0F	TFLG1
$0052	Bit 15	14	13	12	11	10	9	Bit 8	TC0(H)
$0051	Bit 7	6	5	4	3	2	1	Bit 0	TC0(L)

LAB #7:
PIN Verification with 4x4 Matrix Keypad

Example 7.10 in your text is an assembly program to scan a 4×4 matrix keypad and return the ASCII code of the pressed key.[1] Modify this assembly program for the following:

Note: The assembly program in this text is only a routine; you have to write your main program and call up this subroutine for keypad scanning. Example 7.10[2] is a complete routine, and Example 7.7[3] is a simplified version of Example 7.10.

- a. Assign any 4 digits as your master PIN, and push all 4 digits **onto the stack**. Make sure you know how to pull or pop them out from the stack. *Note:* Stack is in first in last out format (FILO).

- b. Insert a 10 ms delay routine into the function found in the assembly program. This 10 ms delay is for debouncing the keypad. *Note:* Your keypad might be good enough to do this without debouncing.

- c. Interface the keypad with your Dragon12-JR. Set up a counter for the 4-digit PIN.

- d. Pull all 4 digits from the stack and compare the 4-digit PIN with the input key. If any pair matches, then the counter will be incremented by 1. If all four digits match, then the counter will be 4. Sound the buzzer to indicate the PIN matching.

- e. Reset the prescale factor in TSCR2 to 8. The calculation is as shown below.

---

[1] Han-Way Huang, *HCS12/9S12 An Introduction to Software and Hardware Interfacing*, 1st ed. (Delmar Cengage Learning, 2003).
[2] Ibid.
[3] Han-Way Huang, *HCS12/9S12 An Introduction to Software and Hardware Interfacing*, 2nd ed. (Delmar Cengage Learning, 2010).

The E clock frequency = 24 MHz,

The number to be added to the TC0 register = 3,750

Prescale factor for TCNT = 8, then $\dfrac{N \times 8}{24{,}000{,}000} = 10\,\text{ms}$; $N = 30{,}000$

Prescale factor for TCNT = 64 = 2^6

$$\dfrac{N \times 64}{24{,}000{,}000} = 10\,\text{ms}$$

LAB #8:
PIN Verification with Single Pole/Common Bus Keypad

a. Assign any 4 digits as your master PIN, and push all 4 digits onto the stack. Make sure you know how to pull or pop them out from the stack.

b. Insert a 10 ms delay routine into the function called found in the assembly program. This 10 ms delay is for debouncing the keypad.

c. Interface the keypad with your Dragon12-JR. Set up a counter for the 4-digit PIN. Pull all 4 digits from the stack and compare the 4-digit PIN with the input key. If any pair matches, then the counter will be incremented by 1. If all four digits match, then the counter will be 4. Sound the buzzer to indicate the PIN matching.

d. Please note that the keypad has 13 pins, and the button can be located by using the chart below. Two general ports have to be utilized for I/O (Port A and B). The MSWORD version of Single Pole/Common Bus Keypad is available on my webpage.

e. With the downloaded Single Pole/Common Bus Keypad program, assemble and load onto the Dragon12-JR development board. Make sure it works as in Lab #7.

f. Modify the program by swapping Port A with Port B.

```
;       PORTB connected to keypad
;
;       PB0 connected to pin 2 (*)      instead of      PA0 connected to pin 2 (*)
;       PB1 connected to pin 3 (7)      instead of      PA1 connected to pin 3 (7)
;       PB2 connected to pin 4 (4)      instead of      PA2 connected to pin 4 (4)
;       PB3 connected to pin 5 (1)      instead of      PA3 connected to pin 5 (1)
;       PB4 connected to pin 6 (0)      instead of      PA4 connected to pin 6 (0)
;       PB5 connected to pin 7 (8)      instead of      PA5 connected to pin 7 (8)
```

;   PB6 connected to pin 8 (5)   instead of   PA6 connected to pin 8 (5)
;   PB7 connected to pin 9 (2)   instead of   PA7 connected to pin 9 (2)

;   PORTA connected to keypad

;   PA0 connected to pin 10 (#)   instead of   PB0 connected to pin 10 (#)
;   PA1 connected to pin 11 (9)   instead of   PB1 connected to pin 11 (9)
;   PA2 connected to pin 12 (6)   instead of   PB2 connected to pin 12 (6)
;   PA3 connected to pin 13 (3)   instead of   PB3 connected to pin 13 (3)
;   PA4 connected to pin 1   *Note:* PA4 is set as output.

g. Reset the prescale factor in TSCR2 to 8. The calculation is as shown below.

The E clock frequency = 24 MHz

If prescale factor for TCNT = 64, then $\dfrac{N \times 64}{24,000,000} = 10$ ms ; $N = 3,750$

If prescale factor for TCNT = 8, then $\dfrac{N \times 8}{24,000,000} = 10$ ms ; $N = 30,000$

12-button keypads

3 × 4 Button	CODES Single Pole/Common Bus										
1	•										•
2		•									•
3			•								•
4				•							•
5					•						•
6						•					•
7							•				•
8								•			•
9									•		•
*									•		•
0										•	•
#										•	•
	5	9	13	4	8	12	3	7	11	2	6	10	1
	TERMINAL LOCATION										

ASCII CHARACTER SET (7-bit code)									
MS Dig. LS Dig.	0	1	2	3	4	5	6	7
0	NUL	DLE	SP	0	@	P	`	p
1	SOH	DC1	!	1	A	Q	a	q
2	STX	DC2	"	2	B	R	b	r
3	ETX	DC3	#	3	C	S	c	s
4	EOT	DC4	$	4	D	T	d	t
5	ENQ	NAK	%	5	E	U	e	u
6	ACK	SYN	&	6	F	V	f	v
7	BEL	ETB	'	7	G	W	g	w
8	BS	CAN	(	8	H	X	h	x
9	HT	EM	)	9	I	Y	i	y
A	LF	SUB	*	:	J	Z	j	z
B	VT	ESC	+	;	K	[	k	{
C	FF	FS	,	<	L	\	l	|
D	CR	GS	-	=	M	]	m	}
E	SO	RS	.	>	N	^	n	~
F	SI	US	/	?	O	_	o	DEL

An assembly program for the Single Pole/Common Bus Keypad is as shown below:

```
#include "Reg9s12.h"
;main program
        org     $2000
        lds     #$2000              ; define stack address
        movb    #$03,PUCR           ; pull up resistor for PORTs A & B
        movb    #$00,DDRA           ; PORT A configured for input
        movb    #$F0,DDRB           ; bits 7~4 output, bits 3~0 input
        jsr     getchar             ; call subroutine
        swi
getchar                             ; subroutine
;scan_star stands for scanning the "*" key
scan_star brset  PORTA,$01,scan_k7  ; is key * pressed?
        jsr     delay10ms           ; debounce key *
        brset   PORTA,$01,scan_k7
        ldaa    #$2A                ; get the ASCII code of *
        rts
;scan_k7 stands for scanning the "7" key
scan_k7 brset   PORTA,$02,scan_k4   ; is key 7 pressed?
        jsr     delay10ms           ; debounce key 7
        brset   PORTA,$02,scan_k4
        ldaa    #$37                ; get the ASCII code of 7
        rts
```

scan_k4	brset	PORTA,$04,scan_k1	; is key 4 pressed?
	jsr	delay10ms	; debounce key 4
	brset	PORTA,$04,scan_k1	
	ldaa	#$34	; get the ASCII code of 4
	rts		
scan_k1	brset	PORTA,$08,scan_k0	; is key 1 pressed?
	jsr	delay10ms	; debounce key 1
	brset	PORTA,$08,scan_k0	
	ldaa	#$31	; get the ASCII code of 1
	rts		
scan_k0	brset	PORTA,$10,scan_k8	; is key 0 pressed?
	jsr	delay10ms	; debounce key 0
	brset	PORTA,$10,scan_k8	
	ldaa	#$30	; get the ASCII code of 0
	rts		
scan_k8	brset	PORTA,$20,scan_k5	; is key 8 pressed?
	jsr	delay10ms	; debounce key 8
	brset	PORTA,$20,scan_k5	
	ldaa	#$38	; get the ASCII code of 8
	rts		
scan_k5	brset	PORTA,$40,scan_k2	; is key 5 pressed?
	jsr	delay10ms	; debounce key 5
	brset	PORTA,$40,scan_k2	
	ldaa	#$35	; get the ASCII code of 5
	rts		
scan_k2	brset	PORTA,$80,scan_klb	; is key 2 pressed?
	jsr	delay10ms	; debounce key 2
	brset	PORTA,$80,scan_klb	
	ldaa	#$32	; get the ASCII code of 2
	rts		

;scan_klb stands for scanning the "#" key

scan_klb	brset	PORTB,$01,scan_k9	; is key # pressed?
	jsr	delay10ms	; debounce key #
	brset	PORTB,$01,scan_k9	
	ldaa	#$23	; get the ASCII code of #
	rts		
scan_k9	brset	PORTB,$02,scan_k6	; is key 9 pressed?
	jsr	delay10ms	; debounce key 9
	brset	PORTB,$02,scan_k6	
	ldaa	#$39	; get the ASCII code of 9
	rts		
scan_k6	brset	PORTB,$04,scan_k3	; is key 6 pressed?
	jsr	delay10ms	; debounce key 6
	brset	PORTB,$04,scan_k3	

```
                ldaa      #$36                    ; get the ASCII code of 6
                rts
scan_k3         brset     PORTB,$08,keyloop       ; is key 3 pressed?
                jsr       delay10ms               ; debounce key 3
                brset     PORTB,$08,keyloop
                ldaa      #$33                    ; get the ASCII code of 3
keyloop jmp     scan_star
                rts
; the following subroutine creates a delay of 10 ms
delay10ms       movb      #$90,TSCR1              ; enable TCNT and fast flags clear
                movb      #$06,TSCR2              ; set prescale factor to 64
                movb      #$01,TIOS               ; enable OC0
                ldd       TCNT
                addd      #3750                   ; start an output compare
                std       TC0                     ; with 10 ms time delay
                brclr     TFLG1,$01,*             ; if equal, branch to itself
                rts
```

LAB #9:
Siren Generation

Modify Example 8.7 in your text to generate a siren between 200 and 1000 Hz.[1]

a. Write a program to generate a two-tone siren continuously that oscillates between 200 Hz for 0.5 second, no oscillation for 0.5 second, 1000 Hz. for 0.5 second, and no oscillation for 0.5 second. Use the buzzer on your Dragon12-JR to output the two-tone siren.

   - Set the prescale factor for TCNT to 8.

   - The delay count for the low frequency is (24,000,000 ÷ 8) ÷ 200 ÷ 2 = 7,500.

   - The delay count for the high frequency is (24,000,000 ÷ 8) ÷ 1000 ÷ 2 = 1,500.

b. Modify your program to generate a two-tone siren continuously that oscillates between 200 Hz for 0.5 second, no oscillation for 0.5 second, 1000 Hz. for 1 second, and no oscillation for 0.5 second. Use the buzzer on your Dragon12-JR to output the two-tone siren.

c. Modify your program to generate a two-tone siren continuously that oscillates between 300 Hz for 0.5 second, no oscillation for 0.5 second, 1000 Hz. for 1 second, and no oscillation for 0.5 second. Use the buzzer on your Dragon12-JR to output the two-tone siren.

The period for a 200 Hz $= \frac{1}{200} = 0.005 \sec$, and $\frac{period}{2} = \frac{1}{200} \times \frac{1}{2} = 0.0025 \sec$.

With a prescale factor = 8, the MCLK is scaled down from 24 MHz to 3 MHz.

Thus the period of each clock cycle $= \frac{1}{3MHz} = \frac{10^{-6}}{3} \sec$.

---

[1] Han-Way Huang, *HCS12/9S12 An Introduction to Software and Hardware Interfacing*, 2nd ed. (Delmar Cengage Learning, 2010).

The delay count for 0.0025 sec = $\dfrac{0.0025}{10^{-6}/3} = \dfrac{0.0025 \times 3}{10^{-6}} = 7{,}500$.

The period for a 1000 Hz = $\dfrac{1}{1000} = 10^{-3}$ sec, and $\dfrac{period}{2} = \dfrac{1}{1000} \times \dfrac{1}{2} = 5 \times 10^{-4}$ sec.

With a prescale factor = 8, the MCLK is scaled down from 24 MHz to 3 MHz.

Thus the period of each clock cycle = $\dfrac{1}{3\,MHz} = \dfrac{10^{-6}}{3}$ sec.

The delay count for 0.0025 sec = $\dfrac{5 \times 10^{-4}}{10^{6}/3} = \dfrac{5 \times 10^{-4} \times 3}{10^{-6}} = 1{,}500$.

LAB #10:
Dim an LED

Assume that we use the PWM0 output to control the brightness of an LED.

Write an assembly program to dim the LED to 10% brightness gradually over ten seconds.

The E clock frequency is 24 MHz. You can set your prescale factor.

Set duty cycle to 100% at the beginning.

Dim the brightness by 10% every second.

Use PWM to dim an LED.

Assume that we use the PWM0 output to control the brightness of an LED.

Write an assembly program to dim the LED to 10% brightness gradually over five seconds.

- The E clock frequency is 24 MHz.
- Set duty cycle to 100% at the beginning.
- Dim the brightness by 10% in the first second and then 20% per second in the following four seconds.
- Select clock A as the clock source.
- Make waveform to start with high level.
- Select 8-bit mode.
- Set clock A prescaler to 64.
- Select left-aligned mode.
- Set period of PWM0 to 0.1 ms.
- Set duty cycle to 100%.
- Enable PWM0 channel.
- Decrement PWMDTY0 by 1 every 100 ms during the first second and decrement PWMDTY0 by 2 every 100 ms in the following four seconds.

Note: Decrementing PWMDTY0 by 1 every 100 ms is equivalent to reducing the duty cycle by 10% in the first second (1 sec = 10 × 100 ms).

LAB #11:
Home Security System

a. Assume that bit PA0 (Channel 0 of Port A) is an input and represents the condition of a door alarm. If it goes LOW, it means that the door is open. Monitor the bit continuously. Whenever it goes LOW, send a HIGH-to-LOW pulse to port PT5 (Channel 5 of Port T) to turn on a buzzer. Provide a program segment to monitor the above status continuously. No keypad is required.

Note: The home security system is using this method to monitor the doors and windows.

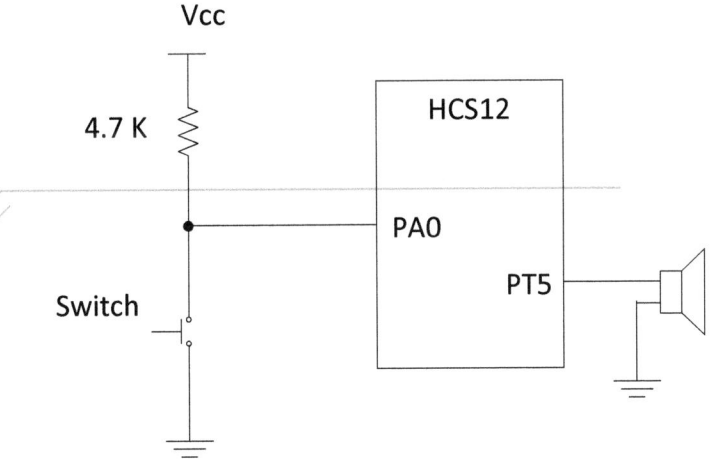

FIGURE L11.1: The magnetic switch provides certain input to the input pin

107

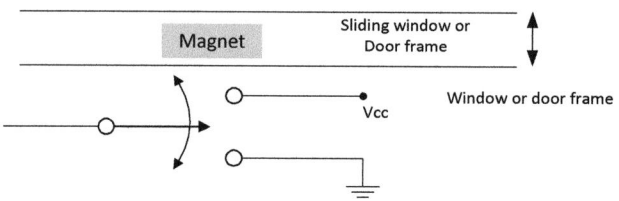

FIGURE L11.2: How a magnetic switch is embedded inside a window or door frame

You can implement the switch by using the jumper wire. When the window is closed, the switch is connected to V_{cc}. When the window is open, the switch is connected to GND.

PORTB	PA7	PA6	PA5	PA4	PA3	PA2	PA1	PA0
PORTT	PT7	PT6	PT5	PT4	PT3	PT2	PT1	PT0

b. Assume that zone 1 for a home security system is to monitor the main entrance PA0 (Channel 0 of Port A), and the door exit to the garage PA1 (Channel 1 of Port A). If **either** PA0 **or** PA1 goes **LOW**, it means that the door is **OPEN**. Monitor these two bits continuously. Whenever it goes LOW, send a HIGH-to-LOW pulse to port PT5 (Channel 5 of Port T) to turn on a buzzer. Provide a program segment to monitor the above status **continuously**. No keypad is required.

Note: The home security system is using this method to monitor the doors and windows.

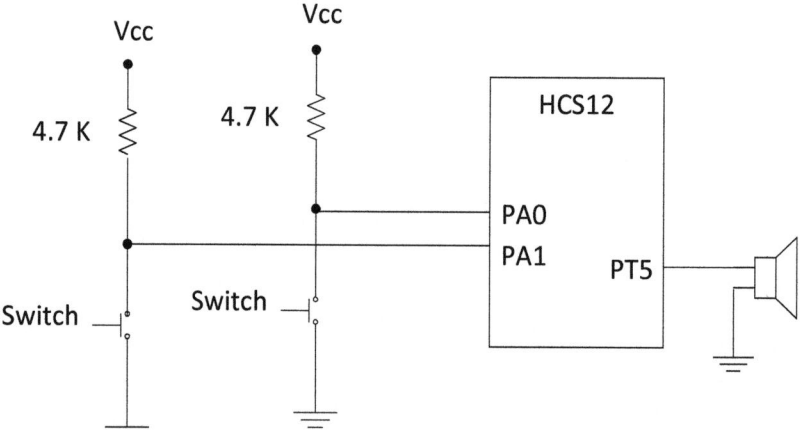

FIGURE L11.3: Each magnetic switch provides certain input for each window or door

PORTB	PA7	PA6	PA5	PA4	PA3	PA2	PA1	PA0
PORTT	PT7	PT6	PT5	PT4	PT3	PT2	PT1	PT0

Printed by Libri Plureos GmbH in Hamburg, Germany